The Technocene

The Technocene

Reflections on Bodies, Minds and Markets

Hermínio Martins

Edited by S. Ravi Rajan
with
Danielle Crawford

ANTHEM PRESS

Anthem Press
An imprint of Wimbledon Publishing Company
www.anthempress.com

This edition first published in UK and USA 2018
by ANTHEM PRESS
75–76 Blackfriars Road, London SE1 8HA, UK
or PO Box 9779, London SW19 7ZG, UK
and
244 Madison Ave #116, New York, NY 10016, USA

British Library Cataloguing-in-Publication Data
A catalogue record for this book is available from the British Library.

Library of Congress Cataloging-in-Publication Data
Names: Martins, Hermínio, author. | Rajan, S. Ravi, editor. |
Crawford, Danielle, editor.
Title: The technocene : reflections on bodies, minds, and markets /
Bb Herminio Martins.
Description: New York: Anthem Press, [2018] |
Includes bibliographical references and index.
Identifiers: LCCN 2018038541 | ISBN 9781783088324 (hardback)
Subjects: LCSH: Social sciences – Philosophy. |
Technology – Social aspects. | Science – Social aspects.
Classification: LCC H61.M29 2018 | DDC 300.1–dc23
LC record available at https://lccn.loc.gov/2018038541

ISBN-13: 978-1-78308-832-4 (Hbk)
ISBN-10: 1-78308-832-X (Hbk)

This title is also available as an e-book.

To Margaret,

and all the decent, upright academics I have known.

CONTENTS

PREFACE

Professor Hermínio Martins (1934–2015) was a leading light in British sociology and worked on a wide range of topics, including sociological theory, the philosophy of the social sciences, the sociology of science, the philosophy and sociology of technology and the historical sociology of Portugal and Brazil. He was widely feted and decorated. The Portuguese government conferred upon him the titles of Grande Oficial da Ordem do Infante Dom Henrique, Portugal (Grand Officer of the Order of Prince Henry the Navigator), and Grande Oficial da Ordem de Santiago e Espada, Portugal (Grand Officer of the Order of St. James), in investitures presided by the presidents of the Portuguese Republic at that time. He also received the Medalha de Mérito Cultural da Câmara Municipal de Cascais (Cultural Merit Medal from the Municipality of Cascais), near Lisbon, and in 2006, the University of Lisbon awarded him an honorary doctorate.

During Hilary term, 1990, while a graduate student at Oxford, I was privileged to attend a lecture series taught by Professor Martins. It was entitled, 'The Theme of Technology in Philosophical Sociology'. It was a survey course, but it encompassed texts, people and ideas that were then, and even now, largely unknown and untaught. The lectures were sparsely attended and, if memory serves me right, only one other student, besides me, came regularly. However, Martins came to every session of the class formally attired in subfusc and, in his thick Portuguese accent, lectured as though there were a roomful of students. The classes were simply an intellectual feast and, without doubt, they formed the highlight of my experience as a graduate student at Oxford.

In the years since, I repeatedly asked Professor Martins about those lectures, and whether he would make the notes available. He always replied that he was working on writing them up as a book. He retired in 2001. We kept in touch casually, and I presented a paper to one of the conferences that produced festschrifts in his honour. However, it was not until mid-2014 that I received, via e-mail, a set of bulky attachments, along with a list of works he had written during the past decade. These papers were not his Oxford lectures, but the contents were gripping. I asked him what he was going to do with them, and it became evident that he simply did not have the strength, in his later years, to see them through to publication as a book. I therefore volunteered to help, and the result is the present volume.

I met and interviewed Professor Martins three times from December 2014 to March 2015. The biographical introduction is a result of these interviews. In editing the manuscript, I made a set of changes after consulting with him. Chapter 1 is a radically altered version of a paper he wrote, exploring the technocene through the human life cycle chronologically. As for other changes, the editorial work has largely been in the nature of clarifying things and, for the most part, I did so in consultation with him. I should add here that this volume is by no means a comprehensive collection of all his works since retirement, or even all his unpublished works. There are other critical essays, in English and Portuguese, on a wide range of topics from art and science to religion. Professor Martins however approved of my selection and was looking forward to seeing the book in print when, without warning, he was taken away from us. Hopefully, in time, a way will be found to make his other writings available to the scholarly audience.[1]

I did the substantial editorial work, but I am indebted to Danielle Crawford, a PhD candidate at the University of California, Santa Cruz, who carefully cross checked references and cleaned up the manuscript. I am also grateful to Helena Jerónimo and José Luís Garcia for their many insights and inputs. Last, but by no means the least, I am particularly indebted to Anthem Press for their patience and encouragement, and to three excellent anonymous reviewers. Although I did not agree with all their observations, I have incorporated many of their suggestions.

S. Ravi Rajan
Santa Cruz, California
April 2018

Note

1. As things stand there are efforts by Professor Martins's colleagues and students in Portugal to collect them in a website dedicated to his works. See footnote 1, on page xvi, for details.

EDITOR'S INTRODUCTION: HERMÍNIO MARTINS AND THE TECHNOCENE

Hermínio Martins was born in Lourenço Marques (now Maputo) in Mozambique in 1934. He was a second-generation Mozambican, in that both his parents were born there as well. Hermínio attended the Liceu Nacional in Maputo, which was then a small city of about 20,000, with blacks, Chinese, Indians, Italians and Germans in a multiracial pot. For a small city, it had a lively intellectual life, with several visiting scientists and artists from around the world giving lectures and concerts. Martins grew up in a household with books. His aunt and uncle, with whom he lived after the passing of his mother, were voracious readers and imported books in Portuguese from Portugal and Brazil, in French from the Continent and in English from a bookshop in Johannesburg, South Africa.

When Hermínio completed secondary school, his family hoped he would attend college somewhere nearby. Geography dictated South Africa, but the young 17-year-old disapproved of the aparthied system there and decided to travel to England, where he knew nobody. He subsequently studied at the London School of Economics (LSE) where, in 1957, he earned a BSc (Econ.), an interdisciplinary degree akin to Oxford's PPE, before doing graduate research under the supervision of Ernest Gellner during 1957 to 1959.[2] Amongst the many highlights of his career at LSE were classes in the philosophy of science with Karl Popper and a deep friendship with Imre Lakatos who apparently claimed that Martins was 'the only sane sociologist he knew!'[3] Martins's time at LSE were the golden years of the discipline of philosophy of science, and these influences were to prompt him to write a classic essay on Thomas Kuhn (Martins 1972).

Martins's first job was at the University of Leeds, where he worked from 1959 to 1964. He taught in the social studies department, with colleagues from the disciplines of sociology, anthropology and political science and, worked, amongst others, with John Rex and Bryan Wilson.[4] Martins recalls several seminars that influenced him while at Leeds. These included what became Piyo Rattansi and James E. McGuire's classic paper on Newton (Rattansi and McGuire 2007) and lectures given by the effervescent genius, Jerome Ravetz. Martins then moved to the University of Essex (1964–71), where he co-founded the Department of Sociology and the School of Comparative Studies. In 1971, Martins moved to the University of Oxford, where he was lecturer of sociology and fellow of St. Antony's College until 2001, when he retired. He was selected especially to teach the sociology of Latin America and remained there before retiring in 2001, when he was elected emeritus fellow of St Antony's College. It might also be worth mentioning here that for a considerable time after the 1974 Carnation Revolution, Martins was a stateless person, in exile in the United Kingdom.

During the course of a distinguished academic career, Martins also held visiting appointments at Harvard University (1966–67) and the University of Pennsylvania (1967–68). He was also widely sought after in the Portuguese-speaking world and was appointed Investigador Coordenador (Research Professor) at the Institute of Social Sciences, University of Lisbon, in 2000–2005 and, since then, an Investigador Honorário (Honorary Research Fellow) of the institute. He also held appointments for shorter periods at various universities in Brazil (UNICAMP, UNESP–Marilia) and in Portugal (Coimbra).

In addition to these formal academic appointments, Martins served as founder-member of the Society for Latin American Studies (UK). During the 1970s, he was convenor of the Theoretical Sociology Group of the British Sociological Association; co-founder of the British Society for Durkheimian Studies; founder-member of the editorial board of the *Sociology of Sciences Yearbook*; founder-member of the Associação Portuguesa de Ciência Política (Portuguese Political Science Association); and member of the Consultative Council of the Associação Portuguesa de Sociologia (Portuguese Sociological Association). In addition, he served on the editorial board of the journal *History of the Human Sciences and Configurações – Revista de Sociologia.*

By the time Martins moved from Essex to take up his Oxford position, he sensed that the discipline of philosophy of science was 'not going anywhere' in the sense that there was 'no great movement', only technical innovations. The sociology of science was by then a thriving new field, driven by radical constructivism that, he felt, was partially a response to the philosophy of science. Martins was, however, dissatisfied with the emergent sociology of science, which he felt was akin to sociological solipsism. The term sociological solipsism, Martins claims, was coined by Talcott Parsons in his discussion of Durkheim (1950), but Martins felt it was equally applicable to the new field of the sociology of science.[5] It was, he held, a doctrine that we cannot know anything beyond society and, in particular, that humanity cannot know nature. This concern, as we will see later, became an important theme in his later work. It might be worth mentioning here that Martins was never a great fan of extremes of radical dichotomies and he disavowed the commonly held idea that two of the great founders of modern sociology, Max Weber and Emile Durkheim, were two parts of the positivist/interpretive divide. Indeed, he went on to edit books on both (Pickering and Martins 1994; Lassman, Velody and Martins 1989).

At Oxford, Martins made two decisions. First, he decided to particularly study modern Brazil, following up on some early research there, starting in 1971. During subsequent years, he travelled extensively there and wrote and published in the Portuguese language on topics related to contemporary Brazilian society. In particular, he explored the authoritarianism implicit in Portuguese–Brazilian relations. He also took on a number of Brazilian students and mentored many young academics, especially from the Global South. Second, his disenchantment with the sociology of science led him to pursue studies in the sociology of technology, in which he became a pioneer. He appreciated the work of the philosopher, Rom Harre; was impressed by the work of two social anthropologists, Edwin Ardener and Rodney Needham; and valued his interactions with Lawrence

Whitehead, a scholar of the politics of Latin America; but no one else at Oxford caught his attention. He, however, read voraciously and widely across the disciplines. He carefully studied the Bronisław Malinowski and Evans Pritchard traditions in anthropology, and invested in close readings of the works of Talcott Parsons, Edmond Husserl, Jacques Ellul, Hans Jonas, Mark Bloc and Hannah Arendt. During his time at the University of Pennsylvania, he met Stanley Milgrim, who, notwithstanding the notoriety associated with his infamous experiment, 'seemed like a nice fellow'![6] Martins wrote prolifically on emerging technological systems, such as information technology, biotechnology and robotics. Martins's first publication in the philosophy of technology was a paper on Ernst Kapp, under the title, 'Hegel, Texas' (Martins 1993).[7] Over the years, he asked probing questions about technology and ethics. He was particularly interested in the question of what technology was a proxy for, and the more classically driven questions, such as 'What is it to be human in the age of modern technology?', and in particular: 'What is it to define ourselves via technologies?' Martins also took note of the wider geopolitical and socio-economic backdrop. By the end of the twentieth century, the social democratic project had largely collapsed and markets had pervaded virtually everything. Thus, his inquiries into technology widened to encompass the interactions between markets and technologies.

Moreover, responding to the trend in the United Kingdom, he began to systematically analyse the marketization of universities and, more generally, of higher education. He believed that members of his own tribe were intrinsically involved in the transformation of an institution he held dear. He pointed out, as an example, that a former sociology colleague went on to become a vice chancellor at a university, only to gain notoriety by cutting social-science jobs and slashing academic salaries. Evidently, the underlying argument for the latter was that a worker's total income ought to be the sum of physical money and psychic income; and because academics performed work that was intrinsically enjoyable, they had a surfeit of psychic income, thereby necessitating the cut in physical income![8] Martins's research also included sociological studies of risk and uncertainty, on human experimentation and eugenics as well as on calamities, drawing upon the work of Sorokin (2010) and on acceleration principles in the social sciences. Martins thus has an impressive oeuvre in the sociology of technology, published in the Portuguese language, and he is widely recognized as a leader in this field in the Portuguese-speaking world.

After retiring from Oxford, Martins also wrote prolifically in the English language. The papers, more than a dozen in number, were substantial, with an average length of more than 25,000 words each, and encompassing a wide range of topics, including the body, mind, information technologies and the cyberworld, in the context of the emergent world of commodities, universities, markets and capitalism. The essays are deep philosophical reflections on our contemporary world and draw extensively and eclectically upon a wide range of theoretical influences including continental philosophy, history and psychology, to name but a few disciplines.

The purpose of the present volume is to publish a selection of these insightful essays. Some of the included papers have been published previously, but either in an abridged form or in the Portuguese language (see acknowledgments).[1] Others had remained unpublished. This book makes these texts available to an English-speaking audience for the first time. Given that I inherited drafts, for the most part, I did have to use some editorial discretion. I have corrected some of the references cited to render them bibliographically accurate and have modified some of the sentence formulations so that they are more accessible – without sacrificing Martins's essential style and, especially, his humour. I did not add new bibliographical references, opting instead to retain the integrity of the original texts. It should be stated that I did manage to discuss my editorial strategy with Martins while he was alive and that he approved of my approach. In particular, we agreed that although Martins's range of writing was broad, encompassing a wide range of ideas and topics, it was important for any particular book to focus on a specific subject. Given that my own expertise is in the area of science, technology and the environment, I decided to focus on essays on these subjects and selected those I believed would provoke considerable thought and debate. I understand that others are working on compiling his other writings and that a comprehensive website is planned.

The works included here all apply theoretical insights, from philosophical sociology to understanding the human condition today. The term *philosophical sociology* is used here in the sense in which Martins intended. It is an attempt at understanding trends in contemporary society through the theoretical lens of continental European speculative philosophy. It is, therefore, a genre that is reflective, evaluative and, in a manner of speaking, normative about rights, wrongs and ethics more broadly. As such, it is quite different from social theory, or the sociology of science or technology or, indeed, science and technology studies. The first essay, which is the only piece I substantially modified, is a painstaking examination of the human life cycle. It explores the various stages of human life, from birth to death, and in the process, examines the intersections between technology and markets in the reconfiguring of gametes, genes, viscera and other body parts. Likewise, it analyses the transformations underway in critical social institutions, including schools and universities, companies and firms, the church, the state, armies and police. Throughout, he ponders the question of what it means to talk about technological agency. The underlying theme is the idea that the 'life trajectories of the human species have been transformed by the concurrent and often interdependent, mutually supportive, technification and marketization of crucial phases of our life trajectories'. The purpose of the first chapter is to examine these interactions between technologies and markets and, in doing so, characterize the key dimensions of what it is to be human at the start of the third millennium, in the age of what he calls the technocene. It is thus a deeply engaging reflection on how the emerging world is totally different from anything that preceded it.

The subsequent essays carry forward these philosophical investigations. Chapters 2 to 4 carefully analyse various implications of the emergent biotechnologies. Chapter 2 examines modern bio and information technologies and emergent ideas about the relationships between the mind, brain and the body. Engaging in a thought experiment, Martins argues that it is conceivable that, in a few decades, technologies might be possible to scan the brain and transfer, without loss, to another vehicle, a process that Martins describes as 'the final technological human discarnation'. This process might well enable the consummation of human 'disenfleshment', and create a neo-biological civilization of androids and the like based on potentially immortal, superintelligent or ultra-intelligent software. However, he argues, such transformations will evoke big, ethical questions. Moreover, the 'decisive questions of real choice' will not lie in the future, but 'right now, if not yesterday'. Along the way, Martins returns to the point at which he departed from the sociology of science, the view of 'the radical social constructivist', according to which 'the body is "nothing but" social constructs and there are no pan-human universals'. This is a position Martins eschews in favour of one in which he urges us to stake positions on the basis of ontological standpoints – and argue for them.

Chapter 3 is about enhancement technologies, eugenics movements and other pathways to the post-human. Here, among many other subtle arguments, Martins implores us to imagine what is at stake in such transformations. He argues that it would shatter the unity of humankind, in both religious and Enlightenment traditions, in the unity of human and nonhuman life and in the human bond with the biosphere. He concedes that humanity has been 'the most predatory and destructive species on Earth'. Yet, he asks us to pause and reflect, warning that 'many vanguards, sometimes wearing the specious mantle of reason and science, of knowledge and power, engaged in a messianic, perfectionist, redemptive mission, have come and gone in the last two centuries or so, their salvific promises exhausted. Not, however, without causing immense damage'.

Chapter 4 continues this discussion but, here, Martins explicitly enters into an argument with approaches to bioethics derived from contemporary analytical philosophy. He offers a pointed critique of conventional bioethics as well as a call to draw moral lines and stake strict metaphysical boundaries. The critique of bioethics is essentially that it 'has lost all capacity to surprise', by becoming a 'a game with preset constraints' in which, with slight qualifications, every biotechnological project is ratified as acceptable. Martins writes that, ethically speaking, this is a deeply flawed and irresponsible approach, arguing that 'a Martian [. . .] would have no sense that Earthlings have been experiencing a series of moral, theological, metaphysical earthquakes owing to the novel, rapid, unsettling character of biotechnological change'. He argues further that humanity has a clear choice to make: either let what he terms the 'Principle of Technological Plenitude' dictate terms, or make a serious attempt to draw boundaries, however arbitrary some boundaries might seem. In particular, his is an argument for the importance of a metaphysics for morals. It would be foolish, he argues, to see the debates about emergent biotechnologies as the clash of two camps 'for science and against science, the enlightened and the obscurantist, the rational and the irrational (fearful, anxious, gloomy, much too ready to predict perverse outcomes and dangerous political utilizations of sound technologies)'. Instead, drawing upon the philosopher, Alfred North Whitehead, he argues: '[T]he mark of a civilized person is to stand unflinchingly for and by some distinctions and boundaries, in issues that he or she cares deeply about, however precarious and challengeable, even if he or she expects to be defeated, if only to bear witness'.

Last, but by no means least, Chapter 5 is a sharp and pointed analysis of the marketization of universities. Alluding to Wittgenstein's idea that to imagine a language is to imagine a form of life, Martins explores the languages, or more broadly, the culture and character of emergent modern universities. He asks: What are the implications of academic capitalism for the forms of life of students? What is the nature of the academic life that they, as well as their teachers, will be participating in? In a wide-ranging discussion of trends and their deeper philosophical implications, Martins examines, amongst other things, the implications for the humanities and the social sciences of what he describes as 'knowledge-as-commodity/knowledge-as-capital industry and the celebrity industry'.

In essence, Hermínio Martins's essays on the technocene engage the points of view and the forms of life underlying the confluence of markets and technologies in the late twentieth century and beyond. This confluence, he forcefully argues, is a giant transformation in the history of both the human species and of the planet. In making his argument, Martins draws upon the entire Western philosophical corpus to address pressing philosophical issues about what it is to be human, where we are headed as a species, and how we need to think, act and be. It is a discussion of the future, but with a deep engagement with traditions of philosophical and religious thought. It is also as strong an argument as one can get for the relevance of philosophical and humanistic reasoning. It is in this emphasis on ethics and spiritual thinking that Martins's contributions to the literature on the technocene is particularly distinctive and important. The purpose of the body of work in this volume is not only to witness and elucidate but to provoke reflective ethical thinking in keeping with Martins's conviction that the social sciences ought to engage in a deep philosophical interrogation of the human condition. This sensibility is excellently captured by one of the anonymous reviewers of this manuscript and, in conclusion, I would like to quote in full a passage from their review, for it does a more eloquent job than I could in explaining the essence of the essays to come:

> This book might be of interest to people who believe that the amazing fantasies about engineering biological and artificial life that have been presented over the past few decades will result in a complete redefinition of what it means to be human. Its focus is on the possible ethical and sociological consequences of an extensive blurring of boundaries between biological and technological aspects of human life. This is a popular field for speculation and imagination, bordering on science fiction. The book might thus appeal to the people who envisage that technological change will overtake biological evolution through biogenetic and technological engineering of humans as well as humanization of machines. People who read science fiction not merely for entertainment, but as a source of visions of the future, are liable to take these scenarios seriously.

Notes

1 Some of Martins's work in Portuguese have been published in two anthologies, Martins, H. (2011), Experimentum Humanum, Civilização Tecnológica e Condição Humana [Experimentum Humanum, Technological Civilisation and Human Condition], Lisboa: Relógio D'Água Editores; Martins, H. (2012), Experimentum Humanum, Civilização Tecnológica e Condição Humana, [Experimentum Humanum, Technological Civilisation and Human Condition], Belo Horizonte: Fino Traço. There have also been two memorial works, Castro, José Esteban, Bridget Fowler, & Luís Gomes (eds.) (2018), Time, Science and the Critique of Technological

Reason: Essays in Honour of Hermínio Martins, Palgrave Macmillan, and J. Príncipe, Évora Studies in the Philosophy and History of Science: In Memoriam Hermínio Martins, Casal de Cambra: Caleidoscópio, 2015.

2 Ernest André Gellner (1925–95) was a famous British philosopher and social anthropologist of Czech descent.

3 The LSE Philosophy of Science department at that time had some of the leading scholars in the field, including Popper, Lakatos, Joseph Agassi and J. O. Wisdom.

4 John Rex (1925–2011) was a British sociologist, of South African descent, who specialized in sociological theory and race studies. Bryan Wilson (1926–2004) was a sociologist of religion, who later taught at Oxford.

5 Talcott Parsons (1902–79) was a leading American sociologist. Martins refers to Parsons's presidential address to the American Sociological Society in 1949 to support this view.

6 Stanley Milgrim (1933–84) was an American social psychologist who, ostensibly influenced by the events of the Holocaust and the trial of Adolf Eichmann, conducted research on obedience. In the infamous experiment, participants were asked to perform onerous acts that conflicted with their own beliefs and conscience.

7 Kapp was a German philosopher of technology and geographer, who was prosecuted for sedition in the late 1840s for publishing an article and subsequently emigrated to central Texas. He developed a philosophy of technology wherein tools were identified as 'organ projections'.

8 During the interview, Martins brushed aside a request for names, stating that his point was not to name and shame an individual but to highlight a trend.

Chapter One

THE TECHNOCENE: ON BODIES, MINDS AND MARKETS

Introduction

Our geological epoch since the 1750s, with the large-scale increase in the use of fossil fuels and thus CO_2 emissions into the atmosphere, has been called the *Anthropocene* by some scientists, including the Nobel Prize winner for chemistry, Paul Crutzen. It could also be called the *Technocene*, inasmuch as the reasons for that denomination, which are because of the impact on the atmosphere of carbon dioxide emissions since the mid-eighteenth century, have more to do proximately with technological agency than with the psychophysiological make-up of *Homo sapiens sapiens*. Besides, while it is not clear whether *Homo sapiens* will survive, there is far more confidence in some circles that technology, in post-human vehicles, will outlive us. Thus, the Anthropocene (in the sense defined) may well represent simply a subset of the Technocene, overlapping for perhaps three centuries or so.

What does it mean to talk about technological agency? Our starting point is the idea that during the past three to four decades, the life trajectories of our species have been transformed by the concurrent, and often interdependent, mutually supportive technification and marketization of crucial phases of our life trajectories, and of most of the key dimensions of what it is to be human. The purpose of this chapter is to examine these interactions between technologies and markets and, in doing so, characterize the Technocene. The chapter has two substantial parts. The first describes the great *transconfigurations* underway as regards the physical body. The second is an examination of the changes in the social institutions that educate, employ, control, regulate and order our lives.

The Body

Birth

We will be born, with ever-increasing frequency, in a hospital or clinic run as a firm (if not in an ambulance of a private enterprise). The medicalization and hospitalization of childbirth has been pretty-well achieved completely, despite fluctuations according to movements of opinion over the last decades, and even longer. Even before birth, we will owe a lot to firms that may deal with our conception in the cases of in vitro fertilization (IVF) – in general under medical control – of uterine insemination and of gestation. With comprehensive medicalization, pregnancy, birth and after-birth care are at the point of being monitored as if they were some kind of pathology. Sexual reproduction is easy and normally absolutely free (one might say market-less and technology-free). However, sex without reproduction is now increasing in the Western world (with an ever-expanding demand for sex toys on the part of both men and women in recent years, not to mention online sex and, eventually, sex with robots), apart from occasional and mostly aborted lapses. With ever-lower fertility rates, the general tendency is for population decline, apart from immigration. On the contrary, reproduction without sex, which is increasing in most Western countries, certainly in the richest, only works with the new reproductive technologies, and the costs in time and money for the parents, biological or other, involved in such dealings can be considerable. The techniques are still quite imperfect, decades after the first successful IVF birth; many cycles of treatment may be needed, and the costs to the would-be parents or to taxpayers have multiplied. Nevertheless, the number of births of this reproduction modality has been increasing steadily in all Western countries and represents an ever-greater proportion of the national total of live births. Fertile couples reduce natural procreation to lower and lower levels (reduced even more by legal or illegal abortions, though legal abortions in hospitals and clinics are now much the greater proportion). In a number of European countries, fertility rates have been below replacement over the last few decades, resulting in declining populations in those countries without immigration, while in ever greater numbers infertile heterosexual couples or homosexual couples resort eagerly to artificial reproduction (medically assisted reproduction).

We may remark parenthetically that the question is more complicated than the mere distinction between fertility and infertility, and not only for medical reasons. There are young, healthy women, between the ages of 18 and 25, the optimal age cohorts for natural procreation, who prefer artificial insemination, turning to sites offering sperm, free or not, by individuals or agencies, because, being economically and professionally independent, they want to have children with the maximum freedom, whenever they want, without sex, without partners, without love, without the vexations of personal relationships. As in many, many other cases, the mere availability of a technology originally designed – or so the scientists involved declared at the time – solely for quite other circumstances (for example the plight of infertile married heterosexual couples) may incentivize the search for and the choice of other options than the normal ones in this area, of sexual love or even mere heterosexual encounters, impersonal and clinical options. The invention of IVF technology was justified at the time, against religious or nonreligious concerns of 'tampering' with the innate human patrimony, as nothing more than a humanitarian response of biomedicine to the anguish and despair of many infertile couples. Infertility affects about 10 to 15 per cent of the population, but the infertile – formerly a silent and ashamed minority – became a vocal minority, exerting constant pressure for the development of technologies and markets answering their needs. But today IVF technology and kindred others also serve single women and even some healthy and fertile couples, free of genetic diseases, who prefer this mode of reproduction – and this may be a rising trend. There are social, cultural, economic and psychological factors at work besides the merely biomedical ones in these options, which are normally ignored or downplayed in the propaganda of beneficent technologies. That the demand for IVF and other new reproductive technologies could arise from factors other than natural infertility and the distress this may cause, was not apparently foreseen, and certainly not publicly affirmed as an additional good (or secondary bad), by the biologists and physicians who developed the technologies in question.

A feminist scholar wrote recently that after the struggle for abortion rights in the last decades of the twentieth century and the beginning of the twenty-first, the next great struggle of women in the area of reproduction will be for the expansion of access to medically assisted procreation, to the 'fertility industry' or the 'fertility market' (Mundy 2007). That implies the industrialization/commercialization of human reproduction on an unprecedented scale. Indeed, it may be that Western societies are evolving towards an unprecedented mode of reproduction, what has been called by another feminist scholar 'the biomedical mode of reproduction' (Thompson 2005). It is not clear whether much of what used to be called 'natural' reproduction, or reproduction tout court, will continue, and perhaps only as a small and diminishing proportion of the total. This is occurring not only owing to the increase in the volume of artificial reproduction (and the selection of embryos prior to implantation via screening for any number of genetic traits according to parental preferences) but also to the increasingly advertised technological possibilities of the 'quality control' of natural reproduction through prenatal genetic screening of the foetus, made ever more sophisticated. Thousands of genetic illnesses can now be screened for and, therefore, the respective genes may be selected against. Some governments have already mandated genetic screening of newborns, doubtless with the best intentions.

Ectogenesis, in the strict or strong sense, with the entire reproductive process of conception, gestation and birth all taking place outside the human body, via artificial wombs, as prophesied in 1923 by the geneticist J. B. S. Haldane, may take some time yet, although it is expected to take no more than a few decades to accomplish.[1] The outsourcing of human reproduction to loci outside the human body, indeed outside any biological body, would involve a societal mutation of sorts, especially if generalized on hygienic/eugenic/emancipatory grounds. 'Emancipatory' at least in the sense of freeing women entirely from reproductive labour (if not yet from child-rearing), and thus contributing significantly to the generalized emancipation of women from the biosocial, physical and normative constraints to which they have been subject in the history of the species. This refers not just to procreative freedom, but freedom from procreation altogether, at any time, for all, permanently, a necessary if not sufficient condition for ensuring a level playing field in competing with males, according to some techno-feminists. The very language used by physicians and other reproductive technologists in these contexts – 'the manufacture of babies', babies as 'products', surely not merely some cryptomnesiac reminiscence of *Brave New World* – is symptomatic of the industrialization in question. As the rate of natural fertility decreases, the investments of parents and firms in artificial fertilization increase, as do even public expenditures when and insofar as the state subsidizes, directly or indirectly, medically assisted reproduction, which it certainly does to some extent already. The public accounts of human reproduction are a singularly neglected area of research, although with the process of marketization of the process of human reproduction it surely should be registered in the national output/income/expenditure statistics, as with other industries. The 'baby market', the 'baby business' as a whole, IVF, the sub-markets such as the market for gametes (sperm and eggs, the latter increasingly flourishing), the hire of wombs for gestation (a truly transnational market now), the selection of extracorporeal embryos via preimplantation diagnostics, and so forth, the project of something like 'designer babies' or the search for the 'perfect baby', are likely to have a great future.[2] Of course, opportunities for non-market transactions may occur, as in the donation of spare embryos as a result of IVF, for this technology typically results in an excess of embryos. However, most embryos arising from this technology never find homes, as it were. We have created a new kind of limbo with this technology, whether or not we classify human embryos as non-persons, though human, or simply as potential humans, or whatever, and thus what duties we owe them. Such surplus embryos, one might argue, are an essentially contested category. The existence of tens, if not hundreds or thousands of frozen embryos awaiting 'disposal' – mostly eventual destruction, occasional donation, especially to infertile relatives, rescuing a happy few from this fate – is certainly one of the unforeseen and unanticipated consequences of the strongly promoted IVF by beneficent researchers. The very existence of extracorporeal embryos, maintained thus for long periods, exemplifies a novel type of entity created by reproductive biomedicine, one which has to be taken into account in ethical discussions: Do they have moral standing (George and Tollefsen 2008)?

Obviously, the new reproductive technologies afford windows of opportunity for negative eugenics, the elimination of embryos for genetic defects of all kinds, even relatively minor or remediable ones, but potentially for thousands of reasons and even, more slowly, for positive eugenics, the selection of the 'best', or even simply 'enhanced', embryos,[3] though one cannot in fact rule out simple parental preference imposed on physicians as of right, according to some bioethicists, with little questioning.[4] We must mention sex selection of embryos on parental demand, now another right recently claimed in the name of the principle of 'reproductive liberty', although not yet formally accepted in Western countries, though strongly endorsed by some Western bioethicists. Moreover, the excess of males in China and India, contrary to demographic normalcy, proves that either selective abortion, gynocide in the womb, after the sex of the foetus has been identified with ultrasound technology, or selective infanticide, after-birth gynocide, have taken place on a very considerable scale. Taking into account the universal entrepreneurialization of the times, perhaps a day will come when the newborn, possibly endowed with the best genetic capital its parents have been able to afford, or with the genetic profile the parents chose, will be registered as a future entrepreneur, a potential start-up, together with an electronic digital birth certificate, which may or may not include gender or other biomarkers, parentage, nationality, and so forth, as the functional equivalent of baptism in the techno-market world they may have to live and die in – or not, if, for example, electronic immortality is bestowed on the masses.

Gametes and Wombs

In the field of human reproduction, the market for gametes, eggs and sperm for artificial human insemination has already been globalized. There are still countries that prohibit the export of such, like India, but none appears to ban their import and, even if they did, international 'reproductive tourism', 'fertility tourism', *tourisme procréatif*, or 'surrogacy tourism' can overcome these statutory barriers (many countries could claim conjoint export–import accounts for these matters). There are price lists provided by specialist firms that may be easily consulted online, although this market is not yet sufficiently differentiated and competitive, with a chronic supply deficit in the case of various ethnic groups in the United States, the UK and possibly also in other European countries despite immigrant communities from other continents. The oocytes of Harvard undergraduates fetch the highest prices by far in the relevant marketplace in the United States (tens of thousands of dollars for each have been quoted), relative to those of women students in other American universities: in fact, in the North American case there appears to be a strong correlation between this list and the international ranking of universities (Almeling 2011). That being the case, it might be cheaper and far less time-consuming to determine the ranking of universities, at least of American ones, inasmuch as this market has not yet made significant inroads in the UK and other European countries by this price list, than by the tedious procedures that have been set up in the last decades, for there is perhaps no better proxy indicator for the purpose.

The choice of sperm donors can be made via videos, or online, according to a range of biometric or psychometric data (height of the male is a prime consideration for women users, often the very top criterion), medical certificates, educational and professional curriculum or the results of Scholastic Aptitude Tests, and photos – the prices being far, far lower and more erratically distributed than that of eggs. Credit scores, a relatively new metric, at least in its systematization by specialist agencies on the basis of online digital data, are now apparently essential for dating amongst certain sections of the population and might possibly be added to the other required data. There are also various paradoxes connected with the new reproductive technologies. For example, an American sperm donor who claimed to be a virgin at the age of 36, a philoprogenitive virgin, the genetic father of perhaps 15 children, was an enthusiastic donor and provided the service absolutely free of charge, advertising online on his own website with forty women clients on his list; he worked freelance, not for an agency. The Food and Drug Administration ordered him to 'cease manufacture' of the sperm on the grounds that he could not provide adequate protection against communicable diseases, a rather specious argument as he pointed out, but the FBI arrested him for that reason, not owing to any complaints from clients or physicians. By law, sperm donations have to be done through reputable agencies. There is, however growth in free sperm donations; buying disease-tested sperm from a clinic can be expensive – possibly two thousand dollars – so 'online underground' websites and chat rooms to connect women with men offering free sperm donations have developed (thus, natural nonsexual insemination). Similarly, of course, there may well be lots of virgin mothers, but if so, they have not received much publicity (though many biologists have long been obsessed with the phenomenon of parthenogenesis), and the case of the virgin father mentioned may be unique. With the freezing of sperm, one can be a father posthumously, possibly with an interval of several years, depending on how successful the cryogenic techniques may be, and of course on the state of the law and on judicial decisions. Some cases have gone through a protracted judicial ordeal to attain the necessary permission for the prospective widowed mother to undergo IVF with her dead husband's frozen sperm, which fortunately was still in good condition at the end of the legal process. Note that the demand for eggs may reach persons of relatively low socioeconomic status; it is not confined to the richer strata. And the transaction can be rather curious, as in the case of an American dancer who sold her eggs in order to pay for breast implants, more or less necessary, she claimed, in her profession (linking thereby two biological markets, both related to her body: the cross-pollination of markets in a single instance).

The market for gametes, even when it must be a black market, has been predominantly North–North, or a domestic one, unlike other biological markets, which are much more transcontinentalized, following an economic gradient, with the buyers predominantly located in the richer countries (as they used to be), as in the case of markets for human organs, genes, gestation or presumably more innocent transnational ones, like that for human hair. Nevertheless, if proposals such as that of legalizing the sale and purchase of human organs were to be implemented, the market for organs at least could become much more national.

Human gestation, or the 'reproductive labour' of women, can be outsourced like other forms of labour, in this case only to other women, for now. The service may be free, or for a price, to relatives (sisters, grandmothers, mothers, aunts, cousins), to friends, or even complete strangers in faraway countries, with all sorts of unconventional implications for the definitions, legal or informal, of family and kinship relations. In principle, the deal could be worked out online person-to-person (or, as they say, peer-to-peer), without intermediaries, as in dating sites. But there already exists a market for the outsourcing of gestation to women in other continents for a price, with possibly additional criteria, with specialist firms handling this delicate business, the prices varying according to the countries concerned and the preferences of the genetic parents, or indeed the prospective legal parents who may not be genetically involved at all, for whatever reason, such as being a same-sex couple.

Surrogacy, or at any rate commercial surrogacy, is banned in France, although there has been some discussion recently about the recognition of the offspring of surrogate mothers born abroad to French genetic parents. It has its defenders, either strictly in favour of noncommercial, voluntary, altruistic surrogate motherhood for friends or relatives, in particular, or even for general noncommercial and commercial surrogacy. For a forceful statement of the latter point of view we may produce the following quote: 'to allow women, who consent in a legal framework, to offer a child to a couple that can not have, would not it be the greatest feminine subversion that one can imagine: finally free from duty to be a mother?' (Mécary 2013).[5] The writer is a lawyer in the Paris Assizes and author of the book *L'amour et la loi* (Mécary 2012). Presumably through a lapse, she writes of the emancipation of women from the duty of being mothers which, in the context of gestational surrogacy, is incorrect, for the women demanding it want to mother/nurse the children that are born of the surrogate womb; the emancipation is solely, in this sort of case, of the presumed duty to gestate one's own children, if health, fertility and marital status permit. Indeed, it does not involve universal emancipation of all women from reproductive labour, since surrogate mothers will be accomplishing the gestation of children for other women.

In the United States, where commercial surrogacy is legal, surrogate mothers may be obliged to fulfil their contractual obligation regardless of how they feel after giving birth. Market–libertarian jurists like Richard Posner have argued strongly that this contract should be legally binding and strictly enforced (1992). India has hosted the largest number of surrogate pregnancies commissioned by Western couples. Money-wise, it has been claimed that it is often a fairly good deal for them or their families (psychic costs and other non-monetary considerations not taken into account). In the UK, commercial surrogate pregnancy is banned, and couples look for surrogate mothers for their genetic children in other countries. It should be noted here that some feminist authors claim that laws banning paid surrogate pregnancy, or the sale of eggs or other fertility services should be rewritten. Women, they argue, should be able to charge lawfully what the market will bear for their work, whatever it may be, and patriarchal laws that try to impede these markets should be revoked (Hakim 2012). This sounds good, but it would imply even greater intensification and extensification of market relations in our societies. *Markt macht frei* is perhaps not a universally valid maxim.

In this field, too, complex questions of jurisprudence arise especially, if not exclusively, when the outsourcing is transnational or transcontinental. Recently, babies born from embryos of biological – or rather, genetic – parents of French nationality, implanted in Indian women as hired wombs, were denied French nationality, in spite of the long-standing rule that the child of French citizens, wherever it was born, was entitled to French nationality. Legal norms in France in this respect have changed significantly in the last hundred years or so. In this specific case, the genetic mother, the egg donor, was anonymous. But many full-fledged biological mothers, in France and elsewhere, who give birth alone and abandon their babies, are unknown; there has been a recent revival in Germany of the kind of medieval institution whereby newborns may be given away anonymously in safe places to be cared for by nuns or other carers. The French Republic, however, remains obsessed with the notion that the womb that gestates a future child of French citizens must itself be of a woman of French citizenship, a kind of reproductive protectionism. The UK, by contrast, is more liberal in this area. Although commercial surrogacy is banned, the offspring of such transactions when brought back to the UK from India or elsewhere, become the legal progeny of British parents with the grant of 'parental orders' by judges, the number of which orders has been increasing yearly.

Cases like this reveal a conflict between the state and the market (even in a fairly market-oriented state) in an area in which the predefinitions of nationality, even according to the *jus sanguinis*,[6] did not envisage or foresee this kind of oddity. One of the typical results of this outsourcing has been called the 'global baby', or the 'world baby', another instance of (human) biological globalization. For it is perfectly possible that the sperm donor, the egg donor, the surrogate mother, the legal father, the legal mother, the birth place of the baby and the country where the birth is registered all be of different nationalities. Thus, five or six nationalities maybe conjointly implied in this tangled process. The legal parents, if of the same nationality, may of course wish to confer their common nationality on the baby with whom they may have no biological, genetic link. A new industry, coordinating prospective parents, donors of gametes, gestating mothers, clinics and hospitals, possibly all located in different countries, has emerged to deal with these multiple involvements. This is however not a static debate. For example, even though at the time of writing (2013) commercial surrogate motherhood continues to be banned in France, a ministerial circular was issued in January 2013 enjoining authorities to issue birth certificates attesting to parenthood and nationality to babies born to surrogate mothers inseminated with sperm from married French citizens, even if it was a commercial transaction.

Questions of this kind will lose their raison d'être[7] when the so-called artificial wombs arrive in the next few decades, with all the functionalities required, and gestation may then take place outside any natural biological body, human or nonhuman, under perfectly controlled conditions (Atlan 2005). Henri Atlan, who makes this argument, also claims that there is a social demand for breeding humans via the artificial womb, presumably over and above those who would rely on surrogate gestation, though it is not clear why they should prefer an artificial womb to other forms of outsourced reproductive labour. Some women would want it, or need it. At the same time, he claims, there will always be some women who want to exercise the privilege, or their natural right, or, in the contemporary parlance, their 'preference', of giving birth themselves. Thus, a world might arise in which a multiplicity of reproductive styles, as it were, would coexist, some of which are not yet technologically available, while others are already present.

Ectogenesis in the strict sense dreamed by the British polymath – biologist, geneticist and evolutionary biologist, J. B. S. Haldane – as the most subversive biological invention he could think of at the time, would at last be a reality. For example, the capacity to ensure the normal development of very premature babies has increased significantly, and foetuses of 24 weeks can now survive thanks to incubators until they are able to survive like normally delivered ones; sometimes this is called an instance of ectogenesis, but this seems to be stretching the term too far. Haldane had announced in the same 1923 manifesto that all biological inventions were potentially subversive and that no social institution whatsoever would long remain immune to the subversions stemming from them. He was in fact referring to what later came to be known as biotechnology, especially human biotechnology, which was given an exalted role as the major revolutionary agency of the twentieth century and the following. We have also been witnessing the realization of some of this potential in the last few decades, and there might well be much more to come in the same subversive fashion. As a socialist, he did not, at the time at least, believe that political agency, the work of political parties and revolutionary elites could surpass it as a revolutionary force, at any rate not as far as the long run is concerned. It would be the climax of the Reproductive Revolution, which has been shaping or at least influencing in major ways, the demography, gender relations, the family and other phases of social life, especially in the West, with the provision of a variety of new technologies that have broken through hallowed and long-standing legal, religious and moral barriers, prejudices and anxieties. We may date this revolutionary process from the introduction of the first effective female contraceptive pill, if not earlier, or from 1978, with the birth of the first test-tube baby (Ferrand 2013). However, the artificial womb, which in a way marks the transition from artificial reproduction to the age of synthetic reproduction, may not in fact be the terminus. The prospect of human reproductive cloning was reaffirmed, after a couple of decades of relative silence on the matter, by a knighted British geneticist,[8] a participant in the 2012 cloning of Dolly. It would be appropriate to accomplish this, he argued, for the sake of bereaved parents of small children. Other geneticists had also pleaded to be allowed to pursue research into human reproductive cloning for the sake of childless, infertile couples. This is of course a contested view. One of the key geneticists in the project that led to Dolly, for example, strongly opposed research into human reproductive cloning. But with or without the availability of human reproductive cloning in the next fifty years, the combination of the new reproductive technologies and genetic engineering or even synthetic biology, may give rise to a 'reprogenetic revolution'.

Genes

Those who are lucky enough to possess gene variants of commercial interest will be able to sell them, or even to set up a commercial genetic/genomic firm for the purpose. As such, firms are now engaged in a world drive to patent everything genetic or biological they can get away with; this mode of possessive capitalism is obviously very contagious. The super-rich, certainly the top 1 per cent of income and wealth in the United States, who have been the target of strictures by the rest, are certainly in a position to buy the best available genes/alleles/super-alleles in the world market for their own enhancement, or that of their immediate progeny, or indeed that of others on whom they may wish, for whatever reason, to bestow their philanthropic largesse in this domain. The anarcho-capitalist philosopher, Robert Nozick, outlined a 'genetic supermarket' on a world scale (1974) – he had in mind human genes, though a supermarket of genes of other forms of life is on the way, and indeed chimerization does rather blur the boundaries between these – to which all parents desirous of enhancing their progeny's genetic capital ought to have free access, at least in legal and moral terms, without interference from states, churches and other authorities. The free market in genes, specifically of super-alleles, has been defended by numerous bioethicists in recent years, a version of the liberalism of biomarkets, or biogenetic market liberalism that has exerted strong influence on Western public opinion. Some indeed have gone further: not only should it be permitted to seek the best or most suitable genes for our children (according to the criteria of the parents) but, in fact, it should be construed as a moral obligation, as a strict duty to seek to do so, if at all financially feasible. The modal shift from *may* to *ought* took place in a couple of decades! Still, they stop short of suggesting that the state should provide capital for the genetic improvement of its entire child population, or provide financial incentives to parents for so doing. But why not? They would become thereby more likely to become successful wealth creators and, despite tax avoidance (at which they would doubtless excel), aggregate tax revenues might increase and economic growth would be stimulated.

Some geneticists or bioethicists are perfectly content to accept the radical implications of this market in a strongly unequal society, where the wealth gap between the billionaire stratum and the rest has increased fairly steadily since the 1970s. They foresee with relish the coming of a reprogenetic revolution (the combination of the new reproductive technologies and genetic ones) leading to the formation of two biological or biogenetic classes of humans, the 'gene-rich' (endowed with the most desirable alleles, 'super-alleles') and the 'gene-poor' (Silver 1997). This could become a two-strata society, and even, according to some who subscribe to this vision, a two-species society, which seems biologically doubtful, as surely they could still interbreed. Still, something like a two-caste society might well emerge in this scenario, with genetic tests being applied to maintain social and occupational boundaries. Human genetic screening has become very sophisticated, cheap, easy and quick to apply, so a genetically transparent society – in which everyone's genetic profile would be, in principle, accessible to everyone – could well emerge, even though the possibilities of genetic modification or enhancement might improve one's chances in the labour or sex markets, where genomic disclosure might be mandatory; one's own genome can be accessed privately at ever-lower prices, with the fast advances in gene sequencing technology and constantly falling costs. That the emergence of such a deep genetic divide between two categories of compatriots should be contemplated with equanimity by top geneticists in democratic America today is noteworthy.

Still, the markets in question, markets in human genes, have not yet developed as had been envisaged by bioethicists. In the last three or four decades, questions concerning the property rights over 'our' genes, the genes resident in 'our' bodies, which were identified and processed by biomedical companies, becoming thereby Intellectual Property, have been the subject of protracted legal disputes in a number of countries, reaching even their supreme courts, albeit with different outcomes in different countries. Some of these cases have arisen from the collection of genetic materials from tribal and ethnic groups without their knowledge or free informed consent. Even if we do not always lose property rights over our genes, our DNA, when they happen to be of interest to a biomedical company, gains the support of governments, some of which have invested politically on a grand scale in this development. The list of patented human genes is likely to grow anyway where such patenting is already allowed. In 2005 there were already more than four thousand human genes patented – that is, about one-fifth of our total genome: more have presumably been added since then, and surely more are to come. Gene patents refer to sequences of genes or segments of genes, or proteins produced by genes: that is to say, the structures, functions and processes of genes. In their great majority, the patents in question do not protect inventions properly so-called, but discoveries of how to sequence genes and related entities. Perhaps for the first time in history the patent system, and the state guarantees involved, have been used to protect the profits of companies accruing from such discoveries, and not from inventions (Koepsell 2009). The genetic matter in question comprises haplotypes, SNP (polymorphisms) and CNV (copy number variation). Besides the patenting of genes, there is an emerging body of jurisprudence in the United States that endorses patents of cells, tissues and even human illnesses. Biomarkets – in an inclusive sense, patents of seeds, microorganisms, multicellular organisms, plants and the genes and genomes of nonhuman organisms – have grown hugely since the early 1970s. There has been not just a gold rush to patent human genes, but a gold rush to patent all sorts of human and nonhuman genes everywhere, and not just genes but also biological materials of varied kinds. The upshot is a very wide marketization of life on Earth in the name of scientific discovery and human benefits by great corporations. The Earth today may be called an 'Agoric Earth', a 'Market Earth', in which everything, especially everything alive, is potentially a resource to be drawn into biomarkets of varied sorts.

The liberal critics of the patent system, which supposedly favour the free market economy (at any rate it is so justified by many vociferous advocates of the free market, if not by the more consistent economic liberals), note that only in this domain have patents stimulated in a very significant way research and development in firms; that is to say on the exploitation of discoveries that should not, in principle, according to the classical theory of patents, legitimately claim property rights. Of course, with genetic modification of organisms, biotechnology firms believe they are on firmer ground to claim patents, but even then they seek long-term control over buyers and users of, for example, their seeds of GMOs, of an unprecedented kind, and use their huge legal resources to the utmost to discipline and intimidate them.

Viscera

Besides our genetic inheritance, we possess blood and organs that may also elicit biomedical interest. Human blood for transfusions has been commercialized for decades in the United States, with significant numbers of destitute people depending on regular blood sales for their livelihoods. The introduction of a market for blood in the UK decreased the voluntary supply of blood donations, which had been very considerable for decades and had been as legitimate a source of national pride as the National Health Service (NHS). The social scientist Richard Titmuss in his classic book of 1971, *The Gift Relationship: Human Blood and Social Policy*, already warned that a crowding-out effect on the voluntary offer of goods or services for altruistic ends was likely to occur with its commodification, and with it a deterioration in quality. This has been much debated, but there is sufficient plausibility for this thesis to provide it with an eponymous accolade as the 'Titmuss effect'. Something like what happened to blood donations has occurred with other varieties of gifts of life, of altruism, following the creation of markets, price scales, and so forth. For example, for one of the major Bolshevik intellectuals, Alexander Bogdanov, a physician by training, asserted the voluntary and free donation of blood, practiced on a very wide, perhaps near-universal scale among adults, would be one of the hallmarks of a socialist society of altruism and solidarity, a gift economy at least in this area (Krementsov 2011). He died officially as a result of a blood transfusion experiment, although it was probably a suicide; he would most likely have been eliminated in one of the Stalin-initiated purges that began sweeping the country not long after Bogdanov's death. In fact, in the USSR blood donations were, in the main, effected in exchange for payment, at least during peacetime. It is interesting to note that the commercialization of human blood donations was a pioneering venture in the setting-up of human biomarkets on an ever-larger scale in our techno-market societies.

It is true that the sale of organs by their 'owners' is still forbidden in the UK, but for how long it is difficult to tell, as many bioethicists insist that it should be unbanned, and left to the discretion of individuals (who of course may be in desperate need of money, owing to calamities or to chronic poverty), but for the time being there is a reasonably flourishing transnational black market in this area. In fact, according to investigative journalist Scott Carney, a close student of these practices, one could speak more comprehensively of a worldwide 'red market'. The 'red market', in his understanding, includes trading in organs, ligaments, hair, bones, blood, embryos, child trafficking and surrogate pregnancy. One of his accounts deals with the Indian 'bone farms', collecting bones from cemeteries, morgues, funeral pyres, and so on, which are then cleaned and processed and then sold for anatomical skeletons used in mostly Western medical schools and labs. There are also 'blood farms' where blood is extracted from poor people for sale to blood banks. Camps of tsunami refugees have been used for the extraction of kidneys on quite a scale (Carney 2011). Curiously, there is also currently a flourishing business in moustache transplants in Istanbul, attracting customers from all over the Middle East. These are elements of the quite varied human–biological globalization processes today taking place at various levels, including of course the microbial one. Needless to say, Carney's case for 'radical transparency' in the supply chain of this market in order to protect human dignity has not been sufficiently heeded.

With the advances of biomedicine, it has become possible to transplant various kinds of organs successfully, and with relative safety for the donors. The simultaneous transplant of five organs in the course of a single medical intervention has already taken place successfully on a human patient. One tends to think of organ transplants as singletons, and of course one has only one heart or liver, but in some cases several organs can be involved in the same transplant procedure. The number of kinds of organs that have been successfully transplanted for varying periods of time has increased quite a lot since the first heart transplant. Recently, faces have been transplanted successfully. The first-hand transplant in the UK took place very recently, and there are as yet only a few such cases in the world. Wombs were transplanted in the UK, at least in one case, from mother to daughter in 2011, but it is not yet certain whether the women who receive such transplants will in fact be able to give birth. It should also be noted that while in some cases donors were dead, in others, they were kept alive just for the operation, after which life-support machines were switched off.

In the next few decades, the range of successfully transplantable organs is likely to increase, making possible in due course something like a modularization of the human body from the standpoint of biomedicine. Modularization, of course, has been one of the constitutive principles of industrialization, quite early in effect in the United States, so much so that a cultural interpretation of the American economy was published with the title *Modular America*, so it is not surprising that the scope of modularization has now embraced the human body. The *Modular Human*, biologically or perhaps better, organologically speaking, would represent something like a biomedical/bio-market version of the *corpus mysticum*,[9] and the globalization of bio-human markets, of reproduction and migrations, would converge into a kind of panmixia, maintaining the biological unity and unicity of our species. It is this biological unity of humankind which some reprogeneticists would like – or at any rate fully expect, in the name of progress and the advancement of the vanguard of the species – to disappear. Yet this unity has grounded the epistemological unity of humankind, from which they, and science as a whole, has benefited and is benefiting now with the radical ecumenization of science.

With the new transplant techniques, and the increasing numbers of those who need transplants for the sake of life and health, new markets have arisen, at least black markets (and there are always entrepreneurs for these activities), though it is not yet clear whether there will be 'white markets' as well with the legalization of the sale of organs. As in the case of other advanced biomedical technologies, such as organ transplants, there arise necessarily thorny ethical issues, dilemmas, aporias, predicaments – at least in contemporary circumstances – addressed in a large body of publications by philosophers, bioethicists, biomedical practitioners and a variety of commentators (religious and secular), economists and other experts. In the future, 'regenerative medicine', the growing of new tissues in particular (the subject of great interest by DARPA, the US Department of Defense's Advanced Research Projects Agency), and electronic or electro-mechanical prostheses, such as the artificial heart, may be commercialized in the next few years; and brain implants for the command of prostheses, and so forth, will surely mitigate, even if they may never preempt completely, the need for transplants. Thus, the ethical issues implicated in organ transplants may lose some of their urgency. At this time, however, the shortage of organs for transplant to match the demand (kidneys, livers, lungs) is driving the search for shorter-term solutions or mitigations, amongst which the procurement of such organs from newly deceased individuals (and, indeed, corpses up to four days old), has gained salience recently.

The rich will be able to buy the best organs for transplant in the global hypermarket for human organs, whether in the black or the white markets (though one might better speak of grey markets in this and other cognate areas). It is not, to be sure, a purely one-way market from South to North or South–South. As market solutions are nearly unanimously recommended by the most influential economists and innumerable think tanks for all areas of life that might be the object of public policies of one sort or another, especially where there are problems of scarcity, the entry of human organs for transplant in a lawful specific market has been defended as the most effective non-coercive measure for raising the supply, if not to match demand completely (for demand has been increasing owing to the Longevity Revolutions). However, voluntary donations, while increasing significantly in recent years, fall far short of the demand. The global free legal market in human organs has been defended by a number of bioethicists, whom one might call 'market bioethicists', or even 'libertarian market bioethicists', since their ethical analyses point prevalently to market solutions for whatever problems they address. They tend to be utilitarian in their ethical approach, though perhaps not all utilitarian or, more broadly, consequentialist ethicists would subscribe to this recommendation of a free market in human organs. A case in point is Robert Nozick, who was not a utilitarian in moral philosophy. Peter Singer, another leading moral philosopher, is one of the leading exponents of 'preference utilitarianism'.[10]

None have defended this thesis with such sustained and extensive argumentation as the American philosopher James Stacey Taylor (2005). Indeed, he goes further than the mere defence of a free legal market in human organs, with every adult being entitled to sell or buy organs, according to their financial circumstances. For him there is nothing immoral in buying organs from those who resort to the sale of their organs in order to survive, or for some other desperate reason or plight. But more than that, he advances the stronger thesis that we have a moral duty, in principle, to participate in the market for organs for transplant. One should be clear that he was not simply saying that we have a moral duty to give organs for transplant, insofar as that would not impair one's health or endanger one's life, if required by kindred or strangers in danger of death, or in other words to participate in the gift economy for such. Many philosophers and bioethicists have argued that given the rising demand for transplants (and states are disposed to incentivize such donations) and the chronic deficit in supply. He was saying much more than this widely accepted thesis. He was saying something much stronger, of a different modal sort: that we have a duty to sell them if we do not want to give organs for transplant voluntarily for free. There is no duty to give, only a permission, for this scholar does not go to the extreme of suggesting a ban on gifts on the ground that markets are always better, given the limiting conditions that one should not thereby compromise one's health and life expectancy (though there will always be an element of risk for the donor as well as for the donee). He emphatically claimed that there is nothing morally wrong in poor desperate people resorting to organ donation. If the thesis advanced by this philosopher were to prevail, and this moral obligation to sell (if not to give) were shared widely, certainly the market for organs for transplant would flourish openly, though how far the supply would increase is moot. Even the bioethicist Lady Warnock, who is certainly not particularly squeamish about biomedical markets, expressed shock and even repugnance at these claims when reviewing the book.[11] It must however be recognized that this book is not so anomalous as all that, for contemporary bioethics is on a path such that more and more theses and proposals will be put forward that will shock even the most avant-garde of bioethicists of earlier generations, who liked to endorse whatever became technologically possible in the sphere of human reproduction, but did not seriously contemplate the ever-expanding marketization of human body parts, wombs, gametes and genes now in process or under discussion. Without using the language of moral obligation, market libertarians have argued strongly, as one might expect, for the legalization and regulation of the market in human organs, especially kidneys, but also livers and lungs, drawing attention to the fact that the market for such organs already exists, on an international scale, but is subject to criminal organizations that are engaging in vicious practices; according to the World Health Organization 10 per cent of all organ transplants are provided by the black market. As one author put it: 'In the U.S., we need a regulated system in which compensation is provided by a third party (government, a charity or insurance) to well-informed, healthy donors. Rewards such as contributions to retirement funds, loan repayments, tuition vouchers for children, and so on, would not attract people who might otherwise donate on the promise of a large sum of instant cash in their pockets' (Satel 2011).

It is the case that, with the considerable numbers of persons who suffer fatal injuries in car accidents yearly, extraction of organs has been taking place with the consent of relatives, if there are any. Governments, bearing in mind this spontaneous, regular, potential supply of organs for transplant, have indeed sought to improve the supply of free organs by requiring as a necessary condition for obtaining a driving license, that the applicant should declare in an appropriate form whether he or she consents, refuses or wishes to consider further the issue of being subject to the extraction of organs for transplant in the case of a fatal accident. The expectation is that through inertia, a majority will simply tick the first box as the default option, and thus their names will be entered on the national register of those who have consented to such a procedure, although an electronic list, of course, is subject to the hazards of all digitally recorded information. It is not clear whether such a declaration is binding for life, as it were, making it an anomaly in these 'flexible' times where all choices, like all prices, can be 'revised', or whether, if the owner of a driving license issued under these circumstances were to change their mind on this issue, they could have this change of mind registered and accepted by the authorities.

A very recent proposal may point the way to a future in which not only the market for organs for transplant will be finally legalized, but it will also be interrelated with other new markets for quite different goods and services (the 'cross-pollination of markets'). The proposal was that English students, whose fees in public universities can reach £9,000 per annum or more, should be allowed to sell a kidney each, at possibly £28,000, a reasonable price, it seems, for buyers and sellers, more or less equivalent to the average annual income in the UK, and which would pay for three years' fees, though not for subsistence during that time, or books and photocopies (Roff 2011). Thus, the burden of the bank debt incurred for the payment of university fees could be alleviated at no cost to the Exchequer, at least directly. Now the broad climate of opinion concerning the sale of organs for transplant has not yet changed so much as to trigger a wave of enthusiasm in the general public for this market solution for the financing of universities through student fees (as well as for the shortage of organs for transplant) as legitimate, necessary or even imperative. Yet the fact that the proposal was made by an academic, a senior sociologist, as it happens a researcher in the sociology of medicine in a Scottish university, is noteworthy. Presumably she made a back-of-the-envelope calculation of the price-elasticity of the supply of organs among needy students. To my knowledge, this is the most striking suggestion to have been reported in the national media, but suggestions have been repeatedly put forward in the UK in recent years along similar lines. Some indicate monetary compensation, however disguised, in addition to travel expenses and replacement of wages lost from time off work, while others offer bribes, such as the promise of free funerals for organ donors, the price of funerals having perhaps increased sufficiently in the UK as to have become a salient bribable item. Ingenious think-tank publicists will no doubt think up other kinds of bribes for this purpose. Oddly enough, I now read in *Le Monde* (4 January 2013, p. 20) that Russian orphans without foster families have been promised by the patriarch of the Russian Orthodox Church free burial sites in the capital's cemeteries, presumably because they have become highly coveted. On the other hand, 'corpse tourism' has emerged in Germany, where the dead may be exported to neighbouring countries with cheaper cremations (Poland, Belgium). Their remains may then stay there permanently.

Regarding the earlier proposal made about university students in England, who are victims of unexpected large rises in fees, there is nothing unique about their plight in economically adverse conditions, except perhaps their age, for they would contemplate a long life without one of their kidneys, for example. Analogous proposals of comparable validity (such as it was) could surely be made, by parity of reasoning, with respect to the indigent, the homeless, chronic dependents on welfare, the long-term unemployed, those in care, perhaps, or indeed any person or category of persons in dire financial straits (many of the same categories of persons once targeted in quite a number of Western countries for measures of negative eugenics).[12]

However, if the supply of organs for transplant go up and up, with a large population of organ sellers, were the supply to rise too much, the price would fall, and the gains to sellers would decrease, and there might be a need to sell yet more organs. One might look at it another way: the estimates of the statistical value of a life, or of our 'net worth' in financial matters, might have to be revised upwards in the light of the proceeds that might accrue from the sale of varied organs of our bodies if such markets are legalized, for these were assets not previously recognized in terms of their exchange value if 'harvested', only their use value for our bodily life, our organic being, our work life. The concept of (human) 'physiological capital' formulated in the 1990s by the economist Robert Fogel (grounded in such bodily attributes as stature, body size, longevity, robustness, capacity of vital organ systems), did not encompass tradable assets but it might perhaps include now 'somatic capital', or perhaps, more exactly, 'organological capital', in the sense of the market value of the sum-total (∑) of all our tradable organs, the organs we might be able to sell in regular legal markets without serious permanent detriment to our health (something of a novelty given the regularity of transplants in the age of biomedicine).

When we are ill, we may be treated in a hospital or clinic, public or private or hybrid but, in any case, we are managed according to market or quasi-market norms or 'disciplines'. We may be unwitting subjects of clinical trials (there is a long history of these in Western countries), but we will be reassured when told that much the greater number of clinical trials for contemporary biomedicine takes place now in far-away countries, on poor and illiterate people. The expansion of markets in open, globalized economies implies the expansion of largely asymmetrical and mostly irreversible outsourcing, a mega-trend of this epoch (though outsourcing is often serial or sequential, moving from country to country, seeking lower and lower labour costs and ever more favourable labour codes and fiscal regimes). The outsourcing and offshoring of industrial production is perhaps the most salient facet of this process, in that much of the manufacturing industrial production of the United States was transferred to China from the 1990s onwards. It should also be noted that there has been some insourcing – or perhaps one should say 'reverse outsourcing' – which in 2012, was much vaunted by Apple, but also followed by firms like General Electric back to the United States. Low labour costs, at least in this case, ceased to be so important compared to other considerations, especially energy costs, which may indicate that the pressures of global labour arbitrage may be abating, but of course not yet on the kind of scale in which outsourcing took place from the United States to China and to some extent South Korea (indeed outsourcing often is a serial process), and without impairing the basic asymmetry of the process. President Obama, in his State of the Union address for 2013, mentioned Apple, Caterpillar, Intel and Ford as American firms that had brought manufacturing back to the United States.

Outsourcing and offshoring occur in many other areas where delocalization is feasible and profitable. Advances in biomedical technology have induced a similar transfer, in this case not too cheap, disciplined and non-unionized labour forces, but to domesticated 'reproductive forces' or cheap and docile clinical subjects, who badly need even the miserable payments they typically receive. If clinical subjects receive miserable payments, revenues for the hospitals, and perks for the doctors who choose or test the subjects in the trials taking place in developing countries, or even in the so-called 'emergent' economies are significant, such as Brazil, India, China, though, as they are already in the top ten economies of the world, the term 'emergent' is a bit misleading). India, in fact, has been a privileged field for these transnational clinical trials, not only for commercial firms that want to shorten the time gap between the lab and the market, but even for nonprofit organizations that also want rapid biomedical advances, shortening the interval between the results obtained in 'wet labs' and their therapeutic application (above all, in the United States), with the prospect of sales everywhere in the world. These trials have been, at least until recently, carried out on young, illiterate, unemployed persons, without the conditions of 'informed and free consent' having been fulfilled, even though the requirement of such conditions has been the fundamental maxim of medical ethics in the West in recent decades, although in practice it is often evaded or derogated in one way or another.[13] Certainly, the local agents involved in such clinical trials, physicians, clinics, hospitals, brokers, officials, and so forth, do not always evince strong humanitarian concerns, their own gains in money or kind being uppermost in their priorities. In sum, it is a poorly regulated market (Petryna 2009).

Even so, the need for clinical trials in an age of great expansion of high-tech research medicine is such that an American professor of bioethics has claimed that there is a prima facie[14] moral obligation on all adult citizens of the country to participate in clinical trials, even against their will, therefore rendering nugatory the much-vaunted maxim of free informed consent – obsolete, according to her. There is, however, still considerable resistance to this move amongst physicians (Cook 2011). If this putative moral obligation were translated into enforced legislation, it could entail a very considerable widening of the universe of 'subjects' of these trials, most adults becoming, as it were, potential biomedical experimentees or trialees, regardless of their own feelings about the matter. We would all, or nearly all if adult, become 'biomedical citizens', so to speak, as those who participated in above-ground nuclear trials in the decades during which they took place; at least owing to the circumstance of propinquity, civilians or military, were called by some scholars 'atomic citizens', or, as those affected physically or psychologically by the Chernobyl catastrophe called themselves, 'biological citizens'. Such a development might reduce the need for outsourcing clinical trials to poorer countries, and to the least able to refuse the meagre pecuniary inducements offered, but the demand for clinical subjects is so huge that not even this beneficial side effect might occur on any significant scale.[15] Nevertheless, even those who subject themselves voluntarily to such trials do not receive compensation if things go wrong, not even for expenses. A few years ago, a very respectable American biomedical institution at the national level urged the recruitment of prisoners for medical trials, with or without their informed consent, with the argument that the need for clinical trial subjects continues to be very great, and therefore it would be appropriate if some ethical or legal restraints were to be waived. They did not dwell on inauspicious precedents: the tradition of subjecting prisoners, especially long-term prisoners, even more so African Americans, to medical trials, including psychosurgery, is a long-enduring one, and not only in the United States.

Stimulation

A related issue concerns 'smart drugs' for cognitive enhancement, especially at exam time, but perhaps also more regularly, since they enhance focussed attention and working-memory capacity. The question of the acceptability of chemical means of cognitive enhancement used by students provoked some controversy, but their permissibility was vigorously defended by leading British bioethicists. Since intellectual work takes place today in extremely competitive environments, with very large populations of research competitors all over the world in many cases, and with competitors often working in better-financed and better-incentivized situations, especially but by no means exclusively in applied fields, there would seem to be an equal case for the permissibility of such drugs for knowledge workers in general. Of course, if their use is generalized, Red Queen effects will necessarily arise, as with all 'positional goods', and relative gains will disappear, requiring further striving with even more potent varieties of cognitive enhancement drugs or other means to match each higher level of competition. Out of the many treadmills of technological civilization that have been diagnosed by sociologists and environmentalists, the best worked out is the 'treadmill of production', though the notion of the 'consumption treadmill' is probably better known (Schnaiberg 1980). Of course, as far as the winner-takes-all rule applies, where the first to discover/invent something or achieve a new record gets all the rewards, there is an incentive to participate. Thus, the records set in competitive sports will of course will be surpassed again and again, year after year, with or without widespread use of performance-enhancement drugs, sometimes by the same individuals over many years and, in many cases, it is not clear how long the fame achieved will last. Actually, the use of illegal performance-enhancement drugs can be a collective practice of whole sports teams (doping collectives, so to speak, perhaps competing with other doping collectives), not just individual sportspersons in quite a variety of sports (cycling, tennis, baseball, running, football, etc.). As one might have expected, some bioethicists have argued that the use of performance-enhancement drugs should be liberalized and controlled only to ensure the health of the users is not adversely affected; the point is the achievement, the conquest of new records, not whether the achievement is solely a result of genetic inheritance and training.

Another related issue concerns 'erotic capital'. The concept has been presented and discussed at length by Catherine Hakim as 'a nebulous but crucial combination of beauty, sex appeal, skills of self-presentation and social skills'. Other scholars, such as R. Baumeister, have also worked out the economic importance of appearance in employment and promotion at work, as well as in sex/marriage markets: 'sexonomics' added to mainstream economics, the law of supply and demand, according to these analysts, applying quite well also to the erotic relations between and within the sexes, in covert or overt, legal or illegal markets of all kinds. A variant would be 'pulchronomics', the economics of beauty (which does not include as yet 'inner beauty'), which is grounded on quite a number of rigorous empirical studies of firms, students' evaluations of teachers and even their election to office in the American Economic Association (Hamermesh 2013). The current boom in cosmetic interventions for body image enhancement in the UK, where it has been claimed its costs are the third-biggest factor of the indebtedness of individuals (a boom preceded by high rates in several other countries, where the demand comes from men as well as from women, if not in such large proportions), demonstrates a certain awareness of these facts. We end up with lots of putative 'capital' modes: economic capital, social capital, and cultural capital being the most commonly discussed, in the wake of Bourdieu's classical if not unique discussions. Economists, besides physical capital, have long worked on the measurement and role of human capital (acquired through education and on-the-job training) and more recently on health capital (the capacity to overcome depreciation in our initial health endowment), Robert Fogel having added 'physiological capital', a concept related to, but distinct from these. In our discussion, we have encountered genetic capital, physiological capital, somatic or organological capital and now erotic capital. Other sorts of capital have been put forward in the business-management literature, such as reputational capital, not to mention knowledge capital. Spiritual capital has been added, not by theologians, but by economists, long after Max Weber's classic writings on the economic ethic of Protestant churches and sects. Of course, nonhuman capital concepts have also been formulated, such as natural capital, ecological capital and bio-capital, which has expanded hugely with the explosion of commercial biotechnology tapping organic matter for fuels, materials, fibres, drugs, food and other applications through biochemistry, genetic engineering and, soon, synthetic biology. The Inclusive Wealth Report of 2012, classified countries in terms of manufactured, natural and human capital. More variants of capital have been proposed regularly in more or less sustained discussions: conversational capital, nonmaterial capital, organizational capital, structural capital, relational capital, emotional capital, political capital, moral capital, religious capital, literary capital, credentialed-skills capital (in recent social-class analysis). The ones already established in the literature comprise more than thirty, and there may well be more to come. Ours is indeed not so much the Age of Capital, as the Age of Manifold Capitals, of pervasive capital formation: capitals in nature, society, culture and personality, as well as in the market economy proper. Put another way, all these realms have been endogenized conceptually in the economy, as the subject of markets, if these capital concepts do indeed illuminate the situation.

Ageing

In the case of ageing and old age-related dysfunctions, there are as yet surprisingly few analyses and recommendations by market libertarians. *Homo senectus* perhaps falls outside the scope of *Homo economicus* proper in analytical terms,[16] not being expected to exercise entrepreneurship, at least in the higher age cohorts, though they are certainly consumers and are major customers of certain industries. They are also voters with perhaps an excessive weight in the 'political market', inasmuch as the young, however verbally political, and prone to engage in street politics and direct action, vote in substantially smaller proportions and thus are in a sense under-represented. We know that the increasing numbers of over 70s, over 80s and over 90s, in Western countries (making up what has been variously called the 'age wave', the 'grey invasion', the 'demographic winter', the approaching 'silver tsunami'), will have no remedy but to ingress eventually in 'care homes' and the like, duly entrepreneurialized as a growing sector of the low-wage, high-cost economy. Given legal constraints and sanctions, and the weight, such as it is, of the ideology of human rights, and the prevalent moral climate in post-Christian societies, insofar as it has not yet been completely superseded by radical individualism, there will remain a standing presumption of a modicum of respect for the human person, but such firms are mostly commercial ones, and thus their interest in us may be in inmates as customers, as sources of revenue. This is certainly one area where the comparative lack of social entrepreneurship projects, of 'social inventions' for coping with such novel and big-scale social problems is very apparent.

Some old-fashioned countries may turn to socio-legal prescriptions, as exemplified by the recent Chinese legislation seeking to enforce with legal sanctions the moral obligation of taking care of elderly parents, for instance by requiring regular visits to them by their children or grandchildren. Perhaps some Western communitarian political think tank may recommend similar legislation. For example, under recent proposals in the first decade of the twenty-first century in the UK the obese will have their benefits cut; smart cards may be introduced to reduce the consumer choice of welfare recipients as the cards will not permit the purchase of alcohol or other undesirable goods; heavy smokers have also been penalized in various ways, and not only via the price of tobacco. Economists, however, like 'libertarian paternalists' (a.k.a. 'soft paternalists') will most likely recommend market incentives, in line with the multitude of incentives provided in recent years by governments for behaviour patterns or decision-making procedures more likely to be conducive to health, longevity, solvency and well-being. Famously recommended in an influential book entitled *Nudge* (Thaler and Sunstein 2008), they have helped to shape modes of state operation not so much of the 'Nanny State' variety (as in health-and-safety regulations in the UK, the absurdities of which have been the object of much denigration by commentators on the Left and the Right), as a complementary 'Nudging State' (I claim copyright for this designation!). So many incentives for this and that, what one might call 'administered incentives' (by analogy with the term 'administered prices', so often the target of criticism by liberal economists in the pre-Thatcher era), that the subject raises serious ethical issues, as indeed also about the discipline of economics construed, as it often is by mainstream economists, as the theory of incentives (Grant 2012). Still, it does not appear that it would be particularly difficult to incentivize appropriate filial behaviour of this kind, but not much in the way of suggestions has come to my notice.

The bulk of proposals, or at least the ones that enjoy wider currency in the West and seem to exert considerable fascination, consist however of technological, or market, rather than social solutions or mitigations. For quite some time now the ideal and indeed universal policy recipe, a standing presumption (at least, in countries like the UK), favoured by governments of varied political persuasions, has been a dual combination – indeed one might say often not just a concurrency, but an imperative coalescence or 'symbiosis'– of a *market fix* and a *technological fix* for whatever social problems may arise, or in connexion with whatever dysfunctions may arise in the received public institutions such as school systems or the NHS. Of course, there are exceptions, in a way paradoxical ones, such as the banning of psychoactive drugs, where the legalization of markets for the most widely used ones is not yet on the agenda of major political parties, though such legalization would accord with the norms they impose elsewhere. In general, this double fix, this double template of a market fix and a technological fix, underlies a great deal of policy thinking in the UK and elsewhere.

In the current technological horizon, the turn to AI software and robotics prevails in the field of technological would-be solutions or mitigations. So, many proposals in the relevant literature point to the introduction of robots of various kinds for various purposes related to the care, protection and assistance to the elderly, though many will be, fortunately, 'elderwells', or in the active pursuit of 'elderwellness'. The most cited types or functionalities of robots in this area are robots for company, robots for nursing assistance, household robots (house cleaning, taking care of clothes, security), robot housekeepers, robot cooks, not to mention self-driving or robot cars to safeguard their mobility. These self-driving cars, if required for everyone, not just for the disabled or the aged, might save tens of thousands of lives every year, not to mention avoidance of injuries and disabilities caused by automobile accidents. To which surely must be added interactive smart robots for conversation, as most humans are too busy to engage in non-instrumental verbal exchanges, and a lot of conversation is in fact quite 'robotic', in the vulgar sense, as once demonstrated by the ELIZA program devised by the humanist computer scientist Joseph Weizenbaum. But great progress has been made since in robotics AI, and we may look forward to the inestimable good of intelligent conversation afforded by such interactions with robots with upgraded AI, complementing or perhaps increasingly replacing those with our fellow humans, perhaps becoming less and less conversable. Or there could also possibly be robots for sexual services. Perhaps 'sexual favours' would be too anthropomorphic a term, at least in the early stages. This is a topic that has been much discussed in other contexts than the human problems arising from the isolation of the aged, such as in PhD theses in respectable universities and in scholarly reports, sex with robots being definitely feasible in their view (teledildonics), though sometimes dismissed as 'vibrators that talk', and certainly commended for public health reasons (avoidance of STDs, so it is claimed). Advances in the manufacture of synthetic skin will provide lifelike tactile gratifications in the human interaction with eroticized robots. As AI advances, and emotional capabilities develop, 'sex bots' might turn into 'love bots', loving as well as being loved (and thus also, presumably, liable to the tragedy of unrequited love).[17] In addition, a vast array of other electronic aids or monitors have been suggested: subcutaneous chips, sensors (in beds, chairs, sofas, etc.) as well as nanosensors inside the bodies, video cameras ensuring permanent two-way contact with the (very) aged, especially the more isolated ones, with physicians, nurses and hospitals or other health agencies ('telehealth'), and of course any number of medical diagnostic apps. In South Korea, for example, the government has affirmed its endeavour to supply by 2020 at least one domestic robot per household, with possibly many functionalities; this would mean millions and millions of robots just in one country! It seems the government has been silent about this project for some time, perhaps owing to the costs or technical considerations. Even the European Union is interested in the development of robots suitable as 'companions for citizens', though it does not seem to be much of a priority, understandably in the current financial conjuncture.

Once again, confronted today with human and social problems of such acuity and on such a scale, the turn to technological solutions, intensive and extensive, in terms of digital or digitally enabled technologies, seems irresistible, on the same lines as the other processes of the computerization and robotization of our lives. For example, many kinds of robots, or extremely versatile ones, would be needed simultaneously to cope with all the requirements of dementia sufferers. Biomedical research on a large scale has been called on to address the horrors of dementia afflicting substantial and increasing numbers in the demographic winter of Europe, though the current biomedical impotence, especially regarding Alzheimer's disease, as yet intractable by widely applicable curative therapies without serious side effects, does not hold much hope for the next decade or so.

A couple of years ago, an American scientist wrote, 'The next phase of sensors – embedded microchips [emphasis in the original], smaller than a grain of sand, implanted into the blood stream by a simple injection. These will monitor the blood constantly and send signals to a smartphone. A ringtone could warn you of an immune response, a heart attack or a cancer that is percolating but not yet a sealed fate' (Goetz 2012). Many futurists have pointed to this development, that is, extended, artificial proprioception, as imminent, or at any rate within the horizon of feasibility within a decade or two at the most. This shows how we will possess an increasingly sophisticated internal technical milieu (as we might call it by analogy with Claude Bernard's 'milieu intérieur' and Durkheim's 'milieu social intérieur', itself inspired by Bernard's coinage),[18] an inside-organism technoculture, as it were, well beyond pacemakers and the like. Technology ceases to be simply 'out there' (of course, strictly speaking, this locution was always misleading: belief in it was a kind of *pons asinorum*[19] in the philosophy of technology) but, insourced, enters within every inner facet of our biological being. We would sense ourselves, especially in pain, not just in the diffuse organic way that has been connate with the species's existence, but also via sensors, that is, electronic microprocessors. Proprioception, emphasized by philosophers such as Whitehead (against Hume, with his exclusive focus on distinct impressions from our outer senses) and Merleau-Ponty (in his vindication of corporeality, against the prevalent mode of disembodied epistemology), will not be the same again.

Death and Beyond

Like our birth, our dying has been hospitalized and medicalized, unless we are very fortunate (for example, dying in our sleep), or very unfortunate (e.g. dying alone and in excruciating pain). Strict protocols may be followed in hospitals in this respect, prescribing the criteria for when to switch off life-support machines, of which we may or may not be aware beforehand or during our stay: moving from medically assisted reproduction to medically assisted termination, or physician-assisted suicide/euthanasia. Dysthanasia may be medical too, at least involuntarily, as medical services of every sort are increasingly robotized and more generally computerized. It might become appropriate to speak of computer-assisted reproduction and computer-assisted termination (if not indeed at least of computer-controlled termination), as ascertaining the readings in electronic screens seems a necessary part of the process of death-determination. The right to die has been proclaimed or demanded in many quarters, justified on the grounds of much protracted and unspeakable suffering. The modal shift from right to duty, which we have noted on some occasions, might well work here: from the right to die, most salient today, to the duty to die, already defended in respectable journals by some bioethicists, who may or may not recall the negative eugenics of the 1930s.

With our death, our corpses may be subject to the sale of our organs, tissues, bones, brains, and so forth, of interest to some hospital, research clinic, laboratory, museum, art gallery, artist or psychopath for medical or forensic science purposes, or others (though even if offered gratis, their extraction and disposal requires the intervention of a specialist company, and thus pecuniary costs will be incurred). In fact, the practical utility, the potential market value, of cadavers, which were normally regarded by the impious as nothing but worthless waste, pure cost in their disposal, has been increasing, thanks to the advances of biomedicine. In fact, some bioethicists have been arguing for the last two decades or more for a legal market in transplantable cadaveric organs. This legally sanctioned market could presumably compensate for the costs incurred in burial and cremation. I am not sure whether the expected market value of transplantable cadaveric organs, given that the demand for organ transplants is so great, far exceeding the supply, and likely to go on rising, could be included in insurance or mortgage calculations, as there is often a cultural lag in these matters. It was discovered recently that some useful kinds of stem cells could be extracted from cadavers up to four days after death; perhaps more such discoveries may come along, raising the biomedical value of cadavers to unprecedented levels. It is not clear whether these flows of utility could be included in market valuations of the total value of humans, at least for the time being, even if there were already a legal market for them or a license for extraction in a living will.

Oddly, given the ubiquity of phrases like culture wars, science wars, religious wars, Darwin wars, time wars, energy wars, the breast cancer wars, the war on drugs, the workplace wars, the war on terror, and so forth and so on, there seems very little talk concerning the clashes (or wars) between 'immortalists' and 'deathists' (believers in the irreversibility of biological death, and the total extinction of the person thereby). These two groups do not by any means coincide with the cleavage between religious and irreligious. Promises of immortality today are as likely to come from secular techno-movements, drawing on the expected advances of biomedicine in the next decade or two, and other areas of technoscience, as from religious movements of old or new provenance, even though they may differ on what they take to be the desirable or possible types of immortality, not to mention on the definitions of mind, spirit, soul, personhood and life itself.

This raises the topic of cryogenics. In the countries where it is already legal (it is – or was – banned in the French Republic), the absolute numbers of those who wait in the appropriate installations is small – but they include sports celebrities, affording a definite propaganda value for the enterprise. Technological alternatives to cryogenics for human body preservation have also been propounded. Such ideas and practices are grounded in the expectation that, with the rapid advances of biomedicine, an eventual resurrection will be possible within two or three decades. The idea is by no means new. During the early years of Bolshevik power, some Communist Party intellectuals hoped for something similar in the case of Lenin. They envisaged quite seriously the possibility that, with the great advances in science and technology that would be instigated by the new socialist order, he could become the first of human immortals. The topic of immortality to be achieved through biomedical and scientific means, either of individuals or societies (in principle, the socialist/communist society when definitively installed), had already been debated among Party intellectuals before the October Revolution, in a cultural environment where questions related to the overcoming of death had been debated in diverse ideological quarters. That is, as long as cryopreservation can be maintained without irreversible damage to the human subjects for the required waiting period, either of whole bodies or just of brains – for, according to one school of thought, there will be a second biological life, in which the conserved and reanimated brain can be endowed with a new body (Masing-Delic 1992).

If cryogenics is not opted for, affordable or permitted, and if there will still be a residue after the extraction of organs, tissues, bones, brains and so forth, the mortal remains can still be dealt with by a commercial crematorium, by far the most successful modality of industrialization ever in the processing of dead bodies. Started in the late nineteenth-century, it now claims the majority of the deceased in Western Europe, in some countries a super majority, practically the default option. Following sound micro-economic principles, the search for the most affordable crematoria for funerals might well be on, and our bodies will migrate from the country of our death (which may also have been the country of our birth) to the neighbouring country with the cheapest crematorial offer (even allowing for transport costs and fees), resting finally there, a terminal local market equilibrium in every sense.

It should however be noted here that there is significant demand for other options than burial in cemeteries or cremation, options of a non-industrial, non-commercial, even ludic kind, in post-Christian European societies, inspired by devotees of ecology, biophilia, humanism, paganism, Buddhism, naturism and so on. Curiously, this is an area where the great expansion of choice and the satisfaction of the most diverse and weirdest preferences, ceaselessly proclaimed in liberal market economies or 'republics of choice', with such slogans as 'procreative liberty' (which could mean keeping gametes congealed for after-death reproduction), 'cognitive liberty' (which includes freedom of access to and use of psychotropic drugs of all kinds but perhaps of little use after death), 'neural diversity', 'morphological freedom' (this perhaps could also justify embalming), 'sexual freedom' (only in virtual posthumous existence), 'freedom of gender' (the last self-definition?) and 'the right to do whatever we want to/with our bodies' (which presumably would encompass self-mutilation, a contagious practice, according to some doctors). It is however not clear whether we could determine what to do with our corpses with the same freedom as we are supposed to exercise with our own bodies. The dead are mostly channelled into just a couple of exits: so much for the exalted value of 'diversity'! Perhaps the invention of new symbolic forms and rituals in matters like these, coupled with alternative modes of corpse disposal, is just too fraught.

The market does not end with death, of course. After death, our posthumous image may be managed by a company, at least for a number of years and perhaps decades post mortem, as is already happening in some cases (mostly having to do with celebrities of the pop-music world). Some firms offer a variety of additional services, especially gratifying to those who would appreciate some kind of eponymous survival, such as assigning your name to a star for the best price (such are the workings of competitive markets). As there are billions and billions of stars, there is scope for everyone, for one-to-one correspondence between the set of personal names and the set of stars, for the entire expected duration of our species. This practice is not endorsed by astronomical learned societies, of course, who claim the sole prerogative of naming astronomical objects for official purposes, but it might one day be a welcome source of revenue if the pursuit of observational astronomy should require funds not available from the usual sources, and governments, under relentless fiscal pressure may well come to encourage it. Recognition of discoveries by amateur astronomers is sometimes accomplished in this way. Posthumous image management for everyone may be a market niche with a future, especially as great philanthropic legacies, the best sort of posthumous image management being the privilege of a small minority.

Our dying can be documented extensively, even exhaustively, should we want it (or even if we do not), and the costs are not too high. The digitized audiovisual documentation of every single second of the dying process might be filed as a *memento mori*,[20] or possibly for the purposes of scientific research in thanatology, the ethnography of life-ends, or for counselling. One might call this type of digital archive a thanatothéque,[21] a personal thanatological library (or document centre, or life-archive). As a caveat, all digital archives are liable to deletion, malicious or otherwise, so back-ups would be recommended, of course. 'Life-logging', the self-documentation of our lives, of everything we do and all our visual or audiovisual images *pari passu*,[22] is already taking place in the pursuits of ordinary people, at least partially, with iPhones, and biometric instruments of various sorts and surely could be amplified almost indefinitely without requiring too much technological ingenuity. We might call 'e-biolibraries' the instances of this type of increasingly fuller and more comprehensive digital archives of human lives. In principle, a digital archive of each individual life, of everything we did, suffered and made, of our praxis, hexis and poesis, could be made. Maybe one day we can also have a record of everything that went on in our brains, with the increasing sophistication of neuroimaging technologies, such as fMRI (functional magnetic resonance imaging). The audiovisual record may be enlarged in due course with the provision of electronic super-senses, and thus also gustatory and olfactory sensors. We will have had our life, the life we led first hand, but also our 'e-life', our simultaneous digitized life (and all our digital acts, as it were, can be accessed by third parties, cyber privacy being something of an oxymoron), which can be filed in the virtual e-biolibrary by a relative, a friend or lover, or a museum/library, perhaps a local one, or stored at a price in some suitable repository, of a kind that may develop as market opportunities should the purpose arise. We already live a good deal of our lives online from very early childhood onwards, especially the hyperconnected among us, who seem to enjoy living in sight of webcams and the like, as much as possible, and increasingly interact with others online audiovisually, including the most significant others, in the most personal connexions, being mostly telepresent to one another (if at all), some of us even amusing ourselves with a Second Life in virtual terms, and define ourselves more and more in computational/informational terms (e.g., the brain, the mind, consciousness, intelligence, etc., as computational entities). Therefore, the transfer of our organic, analogue life to a digital, online, electronic one, as complete a reproduction of it as technologically feasible at a given period, will not amount to such an ontological leap from the real to the virtual. The passage from our increasingly virtualized and augmented reality to Virtual Reality pure and simple may well seem more like a transition in a continuum.

There have already arisen memorial groups via Facebook (and perhaps other social media). In these virtual groups commemorating the deceased, photos and videos of the deceased, podcasts, tapes, recordings of all kinds, e-links, reminiscences and testimonies, personal archives, e-books or e-libraries, are accepted, compiled, and integrated. As cyberspace is infinite, there is room for absolutely everyone, all whom we valued, human and nonhuman, in the record, and there surely is no reason not to include cityscapes, landscapes, cloudscapes (in the original, not the current IT sense of 'clouds'), dreamscapes, and other matters of personal relevance, such as for example the paintings and music the deceased loved. The quasi-divinization of cyberspace can be found in many contemporary authors, as referred to in Margaret Wertheim's book, *The Pearly Gates of Cyberspace* (1999). This re-enacts in some ways the quasi-divinization of physical space in scientists like Newton, who construed cosmic space as part of the sensorium Dei (inspired perhaps by Cabbalistic teachings, which were widely known in learned circles from the sixteenth century onwards), taking Absolute Space, quite independent of time and matter, as a corollary of God's infinitude.

Our entire lives can also be filmed or video-taped continuously, 24/7, minute by minute; the technology is available, and some have already resorted to it, if not always with sufficient determination or persistence (the self-tracking or quantified self- movement is quite significant in the United States and doubtless will be followed in the UK also). The reminiscences and evocations and other memorial acts in these groups may be re-enacted over many years, without the limitations of physical space, or those of the brief liminal occasions, of the funerary and piaculary rites that may still be observed, and of collective memory in general. A kind of quasi-immortality, if not of the body, or of consciousness, at least of the 'virtual soul', an electronic version of what Auguste Comte, whose *Religion of Humanity* revered our dead ancestors (Humanity, the compositum of all human beings that have ever lived, he kept repeating, is mostly composed of the dead), called 'subjective immortality', remembrance and evocations ensured via a great digital social network, and eventually others. Though this is often questioned in conversation, it seems Comte was quite right, and by a very wide margin. According to a recent estimate, one hundred and seven billion people were born in all our history as a species: seven billion are alive today, quite a small fraction.

For those who are the object of one or another memorial group, still not a very large number – even if practically all human dead from now on could in principle enjoy this kind of processing – the most likely eventuality is that, as in life, a stratification may occur according to the status and prestige the deceased enjoyed in their lives, or that which has possibly accrued to them after death, with ratings and rankings by flash mobs, twittermobs or virtual multitudes, which can vary and fluctuate widely, of course, posthumously, like that of authors. Perhaps the immense majority of the dead will remain in a kind of posthumous virtual limbo, as they have always done in predigital days, though virtuality has acquired new meanings with digital computer technology.

Nevertheless, there will undoubtedly arise market opportunities in this respect, as in others – and chances for innovating entrepreneurs – to render posthumous virtual life more widely conferred. This might be seen perhaps as the culmination of the fundamental democratization, reaching out to manifold spheres of life beyond the political realm, and now also posthumous life, on which Karl Mannheim wrote a remarkable paper in the most inauspicious year of 1933.[23]

Social Institutions

Schools, Universities and Firms

Not only is the school supposed to be, or become, part of the enterprise economy, it should, according to increasingly vocal advocates, supply 'enterprise education', training children to become entrepreneurs or at least employable by entrepreneurs, especially in high-tech industries. The enterprise curriculum, according to some self-proclaimed multi-millionaire entrepreneurs in the UK, should be compulsory by law teaching, besides certain academic disciplines still recognized as of indispensable practical value (geography, maths, foreign languages), those skills regarded as most directly contributing to employability. These comprise 'skills' such as honesty, integrity, punctuality, reliability, good appearance, teamwork (Hurley 2011). It is remarkable how honesty and integrity are now presented by businessmen, or their advocates, as skills, while until quite recently they were almost universally acknowledged throughout Western societies (and not only) as virtues, as moral qualities, independently of their possible instrumental interest, even if it might be the case that honesty is the best policy. And of course, good appearance can be code for racism, sexism, weightism and other prejudicial and negative stereotypes. Some of these virtues, like punctuality, application, diligence, reliability, and so forth, were named 'industrial virtues' by the nineteenth century ethicist and historian of morals W. E. H. Lecky, in his great survey of the history of European morals (1869). He used the term *industrial* in the old-fashioned sense of systematic work, not specifically in the sense of manufacturing industry. These were virtues which, in his view, were recognized long before, but became most salient in the modern commercial–industrial economy and society, ranking high in the moral type of the age. He was not, let it be stressed, a moral/ethical relativist.

By stressing these qualities, the ideologues of enterprise education, are, in effect, renaming what were considered virtues as skills, perhaps as a way of justifying the recommendations as purely practical, mundane, nonmoral ones. Moral education, which liberal democratic states have desisted by and large from trying to enforce in publicly funded schools, now re-enters the curriculum by another name. There have been sporadic attempts to revive the teaching in secondary schools of the morale *laïque*,[24] moral education independent of religious commitments, so characteristic of the French Third Republic in the early twentieth century (at one stage associated with Durkheimian perspectives, which in recent years were influential in Normal Schools, that is, teacher-training colleges), the latest in 2011, under the new Socialist administration. Then again, some of these qualities have been presented as 'soft skills', for example the ability to carry out interesting conversations on many subjects, politeness in word and deed, amiability, helpfulness, the ability to be a good listener in certain situations, appropriate demeanour at all times, and so forth, traits that sociologists have characterized as 'impression management' or the 'presentation of the self in everyday life'. A recent report on social mobility, or rather the failure of upward intergenerational social mobility to increase over the last few decades despite the commitment of various governments and the investment in various programmes to that effect in the UK, pointed to the failure of state schools to provide education in such skills, unlike public schools. It is worth remarking here that there is an analogy between skills (*technai*) and virtues in Plato; it is, however, an analogy and not an identity, neither category being subsumed under the other. A lengthy discussion of this analogy in Plato can be found in a recent book by the philosopher Julia Annas (2011).

Moving to the next stage in the educational process, the university or other tertiary education organizations, are also entrepreneurialized in some fashion, formally or otherwise. The classical model of the public university has been undergoing successive attacks by governments of various persuasions in Western countries, with the UK in the vanguard, both in practice and in ideological terms. The universities, not least the research universities, are being ever more closely integrated into the market economy, governed as if they were ordinary market organizations, assimilated to the paradigm firms/enterprises of production (vice-chancellors or directors of institutions such as the London School of Economics being called informally, but very widely, CEOs, until such a time as it becomes the official designation), induced to become ever more completely engines of the 'knowledge industry' (though of course much of the work of this industry, if not the bulk, takes place outside the universities, in the research or R&D arms of openly commercial or industrial organizations), 'cognitive capitalism' or the 'knowledge society'. It is not the scholar, not the sage, not the thinker, not the teacher, not the discoverer, not the inventor, not the researcher per se – however superb in their vocations – who receives official commendation, but the 'academic entrepreneur', although the figure of the (non-university) scientist–entrepreneur–billionaire exemplified by J. Craig Venter of genomic fame, may be even more exalted.

According to some scientists, universities could live from patents, with scientific research being directed towards subjects and topics most remunerative from this point of view, which is happening in any case, without any hope of foregoing other, more substantial contributions to the university costs, as the public universities used to enjoy. For the time being, governments are contributing substantially to the STEM (Science, Technology, Engineering, Mathematics) disciplines, with important contributions from the military in the case of the United States. But the teaching of the humanities (foreign languages and civilizations, classics, history, philosophy, etc.), except as auxiliaries to war, propaganda, control of foreign populations in occupied or 'protected' areas, surveillance, intelligence, cyberwarfare, national security issues of all kind, and the like, the social sciences (with some exceptions of a more mathematical or statistical kind, or for the purposes just listed) and the arts (renamed 'creative industries'), is likely to be left increasingly to market forces, entrepreneurial ingenuity and consumer volatility. True, some have coined the acronym STEAM, for Science, Technology. Engineering, Arts and Mathematics, to liberalize this phrase, but this usage remains quaint. In the main, these developments will mean that without the backing of public funds, the customer or the donors of financial contributions may well shape the kind of research being done, even though 'crowd-funded' finance of science projects by citizens and citizen groups, which has only just started, at least under this description, may well eventually come to play a significant role and establish a more fruitful dialogue between scientists and citizens than has been the case hitherto, with the militarization and the commercialization of so much scientific research since the 1940s, as part of a new form of science-making: 'networked science'. For of course there were remarkable comparable earlier cases, such as support by the general public and voluntary associations to raise funds for and generally assist biomedical research into a vaccine for polio, which campaign lasted for decades, until it reached a successful resolution and two vaccines began to be widely available (Sills 1957).

Sadly, even some young academics today are satisfied with the deepest, the most radical, and ultimate mode of marketization of the university: the redefinition of the relationship between teacher and student into an ordinary commercial relationship, between a provider (seller) of services and the customer (or consumer, financed somehow, buying those services, like any others). If this tendency continues, it will suffice in due course to drastically transform the ethos of the university; there will be no one to recall that things were once quite different.

At least one academic philosopher of distinction has argued that philosophy could and should be envisaged as a form of engineering, although of what is not entirely clear. It is true that in the twentieth century a number of engineers became philosophers, even famous philosophers, but did not profess to do engineering when they were doing philosophy and hardly made any reference to engineering in their philosophical texts. There may well be a rush soon to rename every academic discipline as a mode of engineering, a term which is coming to have in late capitalist countries the kind of prestige it once enjoyed in the USSR, where even writers were commended as 'engineers', too, if 'engineers of the soul' (even though the 'soul' was deemed fictitious), though today in Russia they might be officially redefined 'entrepreneurs of the soul', the secular equivalent of the American 'pastorpreneurs' (the word was coined some time ago, as appropriate for culture that exalts entrepreneurs and entrepreneurship). However, social engineering remains a pejorative expression in the UK and the United States (though the practice is of course engaged in, being designated by other names), which rather impedes the conversion of the social sciences into variants of social engineering, much to the discomfiture of some practitioners who would be only too proud to be so envisaged and rewarded. By contrast, 'financial engineering' is, in the main, an okay expression, and the practice has certainly flourished mightily in the last couple of decades. The coining of the expression genetic engineering was a marvel of publicity (it is a kind of engineering that has no place in Faculties of Engineering, however Catholic). The coining of the term software engineering arose in a context of protracted struggles between corporate management and technicians in the information and computation technology industry, and to this day computer software programming resists automation, retaining a vexing kind of intractability from the point of view of management. 'Imagineering' seems to be a more or less neutral term. Geoengineering may come to attract opprobrium in some quarters. Molecular engineering has not stayed the course, being replaced by nanotechnology (though molecular technology is still used, with less prominence, and the Molecular Revolution arising from the development of molecular sciences, with molecular biology in the lead, is still acclaimed). 'Re-engineering' is much discussed in management literature and is now used well beyond that domain (management-speak does get widely diffused).

Scientific work will increasingly be done in firms (institutes, universities laboratories redesigned as firms or closely associated with firms, formally and informally), or at the service of other firms, in a global market of science business, science for sale, privatized science or commodified academic research. Many scientists, even among the most 'pure' ones, like Einstein, applied for patents for various inventions (to little advantage, in effect, in his case), but such activities took place as if they were subsidiary or even peripheral to their basic interests as scientists; mostly they neither gained much profit, prestige, status or influence from such sidelines; there were notable exceptions of course, especially in Germany, as the cases of Haber and Nernst remind us. But today the typus of the free, independent, autonomous scientist (the very paradigm of the inner-directed personality) tends to yield to the more prevalent – or at any rate more congruent with the spirit of the age – figure of the scientist–entrepreneur or even the triple figure of the scientist–engineer–entrepreneur, of which there are many examples in Silicon Valley and other similar configurations in the United States and other countries (some of which have sought avidly to emulate Silicon Valley).

In the last few decades scientist-entrepreneurs have made up an important contingent of start-ups in areas such as the new technologies of information and communication. But many have been playing important roles in various fields of biogenetics, a key example being the aforementioned Venter, also a genius in self-publicity as well as a billionaire, whose private venture, driving at maximum speed, was able to surpass in some respects the publicly funded Human Genome Project (which for political reasons had already been conceived as a 'fast' project). Venter became one of the founders of synthetic biology, more radical than genetic engineering in its common acceptation, though he himself is being overtaken by even more radical geneticists, some of whom now expect the age-old dream of the laboratory manufacture of a living cell within decades, a 'Second Genesis', a 'Second Creation', or a 'Regenesis', as they call it already (with many other achievements of biomedical value long before that epochal event). Numerous scientists who do their research in private or public universities have strong links with firms, not least big corporations, in areas of genetics or biomedicine, or are co-founders of firms in these domains. Even if not scientist–entrepreneurs, many scientists now own shares in the firms for whom they have done research, for they were paid in part with shares of these firms. Especially in the case of biomedical firms of the pharmaceutical industry, of Big Pharma, the rise of the scientist–shareholder is a cause for concern, given the obvious possibility of a conflict of interest in having a stake in firms for which they research. At the limit, we would have not only a 'capitalism of shareholders', in which the increase of share values becomes a prime force in the direction of enterprises, but a 'science of shareholders', in which scientist–entrepreneurs or scientist–shareholders participate in this variant of capitalism. Besides shares, scientists often benefit from stock options, which bind them more closely to the interests of big firms, especially in the pharmaceutical field. Even essentially salaried researchers in academia work for institutions that are increasingly commercialized or entrepreneurialized – in practice and in their ethos – and dependent on the practices and expectations of the capitalism of shareholders, globally speaking, whatever their subjective sense of inner independence in relation to the world of 'total capitalism'. In any case, the emergence of the 'science of shareholder scientists', as we characterized it, was not foreseen by the scholars in the social studies of science, immersed in laboratory studies as microcosms of science work.

Besides their entrepreneurial roles, many scientists are also managers in universities, not least in connexion with their teams, or in research centres, out of necessity or from a sense of vocation. Many more will have to follow courses in management and marketing as so much scientific work is done in teams, and sets of teams, transnationally and of often-long duration (measured in years or even decades). There are also collective projects associated with heavy capital investments in big machines such as the Large Hadron Collider (LHC), the biggest science machine of all time. The collectivization of physics research has long been established, resulting in many articles signed by well over one hundred 'authors', listed in the alphabetical order of their surnames. Comparable collectivization of research and, hence, publication is now being demanded in the social sciences which, however, lack the large laboratories and experimental facilities enabling series of experiments taking place over decades and which have stimulated such occurrences. All in all, the fascination of the 'hive mind' is growing. But, no matter, management-speak and market-speak tend to pervade the fields of research science and academia, turning our world into at least an apparent cultural Homogenocene, comparable to what scientists have indicated for the biological world (*Homogenocene* is the very word they have used) as a result of the ongoing reduction of biodiversity, the sixth and fastest extinction in the history of life, brought about by our anthropogenic, technogenic and market-led practices, thus the first without violent telluric or asteroidal causes, the first to be induced by one species, the imperial species.

Enterprise education in school and university serves two purposes: to supply entrepreneurs and to supply employees. In other contexts, employability is the key term, in the context of devaluing 'academic' interests and values, stressing above all practical concerns as to what education is mostly about, or should be. Nevertheless, the teaching of maths is not 'practical' in the same sense as the teaching of many skills, unless perhaps reduced to such things as commercial arithmetic or the rudiments of accounting, though it is true that most skills, of every kind, are undergoing technicization of various sorts, either replaced by software algorithms or enhanced technologically via prostheses and implants or through coaching, though often dubious. As 'flexibility', or the expectation that no one will have secure, stable employment in the future (and certainly not for a professional lifetime) is now taken for granted, one would expect publicly funded schools to be enjoined to educate for this flexibility, though this *desideratum*[25] oddly does not appear in any list published by the entrepreneurial propaganda associations. Yet in other contexts market triumphalism is accompanied by praise for the Permanent Gale of Creative Destruction of techno–economic innovation. Indeed, one of the most powerful epithets in the armoury of 'Washington consensus' economists and think tankers is 'rent-seeking' (under which protectionism, corporatism, monopoly practices, and so forth may be subsumed). Of course, any search for security, for stability, for continuity, for the avoidance of instant obsolescence, can easily be so characterized. Only instant, unresisting, preferably enthusiastic, compliance with whatever economic changes supervene, no matter how calamitous, can escape the charge. This of course entails constant changes in occupational structures – geographical, even transcontinental, displacements of industries, ever-shorter product life cycles, and so forth – and thus constant dismissals and, at best, constant retraining in public organizations. The constant large-scale reorganizations of the NHS, for example, are notorious, not directly owing to technological change so much as to political interventionism. We are more conscious than ever now that techno–economic innovation also comprises innovation in ever-more abstruse and incomprehensible financial instruments and packages thereof. The financial services – or disservices – 'industry' is a key branch of economic activity, and the one with by far the fastest processing of market signals of all industries (processing increasingly carried out by computer software algorithms), which now attends ever more closely all other industries as its less-visible partner, monitor, controller, destabilizer – including agriculture, which was relatively late in its absorption into the global financialization process.

Some years ago, an estimate was published predicting that UK graduates could expect to experience five significant changes in occupation during their working lives, perhaps as serial entrepreneurs with their highs and lows, possibly serial bankruptcies, or, as serial employees, with concomitant shifts in industry, rank, income, and so on. In their personal lives, similarly, mutability will become the norm, statistically or even in value/normative terms. They will expect to change locality, country of residence, partners, friends, nationality (dual nationality/citizenship being now accepted by a number of countries and enjoyed by millions), religious affiliation or disaffiliation. They are entitled to any number of sexualities – the number of legally distinguishable sexualities, well-defined or indeterminate, according to some authorities, being very considerable, more than the six or seven that come easily to mind, and individuals may shift repeatedly along the spectrum. They are also entitled to the 'freedom of gender' that is now much easier legally and more acceptable socially, given that declared genderlessness or 'prefer not to answer' are legitimate choices. Appearance may even be changed drastically, zoomorphically for instance, in the more radical variants, as expressions of 'morphological freedom', including self-administered 'body hacking', which can be quite extensive and innovative, not to mention the comparatively milder, repeated 'procedures' of cosmetic or aesthetic surgery of one sort or another, which are only partly expressions of the claim: 'We have the right to do whatever we like with our bodies'. Quite often, this is part of the search for increased 'erotic capital' in labour and sex markets and for improved self-esteem. There were over 950,000 such procedures in Brazil last year, mostly for non-clinical reasons, the second country in the world for such, next only to the United States, and it is not just a question of numbers; it has been claimed that in the UK the cost is the third cause of indebtedness of women. There will therefore probably occur three million or so each year, though significant numbers will be repeats.

An important consequence is that their psychic profile is also more mutable. 'Protean selves' (Lifton 1993) emerge (or define themselves as such): reinventing oneself more than once may become the standard expectation. In epochs of accelerated transformation many people can experience successive self-definitions or reinventions. According to one observer of China today:

> Each time I go back to China, I prepare myself to hear stories of surprising ways that the country has changed since I was last there. I also expect to hear equally surprising reports of personal transformations. It seems inevitable that at least a couple of friends whose lives seemed set to move in one direction will have had something completely unexpected happen to them since I last saw them. During the months since my previous trip to China, a professor who had no interest in business will have become an entrepreneur. A loyal bureaucrat within the Communist Party Youth League will have turned consultant to an international corporation (and now be a bit embarrassed by his earlier commitment to Marxist economics). Someone who considered all forms of religion mere superstition will have become first a fervent Buddhist and then a devout Christian. An earnest graduate student who once said that a visit to Hong Kong was probably the closest that she would ever get to going 'abroad' will now regularly be taking trips to Europe.
>
> China has become what the United States famously was a century ago, a land of reinvention. Rags-to-riches stories are as popular there now as they were in America in the days of Horatio Alger – and for similar reasons. I often feel that many of the people I know in China have lived out several lives while I have been making my way through just this one. (Wasserstrom 2012)

These mobilities and mutabilities, or shall we say 'flexibilities', are facilitated by widening ranges of object-choice and self-choice made available by techno-market forces (though one may be 'forced to be free' to choose, in such instances), such as online screening or tele-work, though the circumstances of the precariat – no longer a small minority of the 'economically active population' in the West but rather approaching the normal condition of everyone, bar a small minority – hardly conducive to the optimum conditions of free choice.

On average, during his or her lifetime an American changes jobs more than ten times, moves house more than six times and marries more than once, the number of changes of non-spousal partner (once sampled through the occurrence of palimony), personal creeds, religious or political dispositions and so forth, are not reliably estimated by or inferable from the available statistics (Gilbert 2006, 214). Not just diachronically, but even synchronically, they resort, out of necessity mostly (though even then there may be secondary gratification arising out of it), to a bundle or portfolio of multiple concurrent jobs, or firms, or economic or non-economic activities of various sorts, bearing some analogy to agrarian economies, with such examples as worker–peasants or peasant–worker–artisans (once famously analysed in certain areas of Northern Italy). In their personal lives, too, concurrency may prevail, in this area perhaps more as a matter of choice rather than necessity, as in the case of polyamorists, for example, not just serial monogamists, as has been the case for decades, the expression 'serial monogamy' emerging in American sociology of the family six decades ago, to depict the realities of marriage in a society still committed by law and custom to monogamy-at-a-time, legal or semi-legal concurrent polygyny being confined to tiny segments of the population, but practicing sequential polygamy. With a divorce rate of about 50 per cent in Portugal today, sequential polygamy, whether successive marriages or partnerships of a few years' duration, will increase considerably. Similar trends, though with lower rates of divorce, are in play in a number of other Western European countries. We have not yet become fully conscious of our North American/Western European societies as societies of polygamists, or at any rate approaching that state (in the sense of serial or sequential polygamy, with the observance of synchronic monogamy). In Western Europe, this coexists with the presence of enclaves of internally lawful polygyny owing to the presence of Muslim communities, a small number of believers actually practicing it.

Firms

If a university graduate looks for a job and is successful, she or he will most likely work for a firm (or several concurrently), as public-sector employment is in steady decline. In any case public-sector organizations aim at mirroring the practices and criteria of firms in markets. Firms, in general, operate in a shifting and insecure market, or rather outer market (for their internal markets are subject to volatility and arbitrariness) and, indeed, often in a world market undergoing the pressures of labour arbitrage, thus the likelihood of outsourcing/offshoring to other countries, to automatic devices or simply to computer software algorithms. Indeed, there are now algorithms for sports journalism, authorship of romances, student essays, grading exam answers, writing technical reports, peer reviewing of biomedical articles, scientific research (the so-called robot scientists), or for trading in stock exchanges, the 'high-frequency traders', replacing human traders on an ever-increasing scale (as recommended by semi-official reports on the financial services industries), and so forth. One can freelance of course, and with the casualization or fragmentation of labour in many branches of economic activity this may be the result, if not the preferred option.

By 'fragmentation' of labour I have in mind particularly the recent development of 'microtasking' for firms like Amazon and other 'digital sweatshops', as they have been called by some critics. Microtasking works by outsourcing small, virtual tasks to an army of online workers, who then perform them for pennies. These tasks vary widely in scope and substance, but what links them all is that they are essentially too difficult or too dependent on human analysis for a computer to do, but too simple for skilled labour. And they are the bedrock of the Internet. Crowdsourced microtasking – conducted largely via Amazon.com's Mechanical Turk site – is now a multimillion-dollar industry, and one that does not appear to be slowing down anytime soon. Even as the global economy continues to falter, Turk is thriving, in no small part because of what it can do for companies under pressure to do more with less (Cushing 2013).[26] IT industries have not been by any means lacked fairly systematic attempts at deskilling, or ensuring the subordination of technical staff, but software designers or engineers have remained unmanageable by the firms that hire such people, despite repeated endeavours over the last fifty years to bring their employment conditions into line with the rest of the white-collar labour force. In fact, a counter-movement has occurred in the last few years, as 'code stylists' and top programmers have become significant entrepreneurs in their own right.

States have been encouraging the 'spirit of enterprise', if not always providing the means for ensuring the capital help for enterprises; nevertheless, any of us can register as a firm, or several at the same time (at least in the UK), possibly including all adult members of the family/household. In some cases, children, such as child models or teen singers or teen hackers, can be added to the entrepreneurial pool, as it were, as formerly they might be added to the labour pool of the family as a collection of wage earners. As self-employed, we can be employer, employee and entrepreneur concurrently. We can glorify our condition by insisting on being called 'entrepreneur': one could say at the same time, but perhaps at different times of the day one could take turns at the different roles, the distinctions arising more from analytical or conceptual considerations rather than from empirical ones. Indeed, the expression 'self-entrepreneur' was coined recently as one who runs a 'self-enterprise'.

Given the pervasive pressures for redefining all occupational or professional roles as entrepreneurial, as if this were the only kind of excellence, it was only to be expected that socialization for entrepreneurship is being formally developed. There are now nursery schools where the merits of entrepreneurship are taught and entrepreneurs in national and international life glorified – the 'heroes of production', or any rate the 'heroes of entrepreneurship', or the 'heroes of finance', positive or negative, are back, even if the heroes of consumption, in Leo Lowenthal's classic formulation (1984), the media celebrities, continue to preoccupy the media and vast tracts of cyberspace, though we should perhaps include also the 'oligarchs' and all those who can afford ever more grandiose superyachts. The Girl Scouts of America, for example, organize competitions for the presentation of the best 'business models'. Just as in some Catholic cultures families and the priesthood sought to stimulate in the whole populace from early childhood the maximum of vocations for the religious life, for the secular or regular clergy, the religious orders for men and women, in this brave new world, incentives are being set up to stimulate the maximum of vocations for entrepreneurial life or, failing that, at least for the role of good customers, of good consumers, from quite early in life. The socialization of even small children for the role of consumers/customers in a market economy now begins quite early. There are apps for iPhones for children from the age of three months onwards (touch becoming, as it were, the initiatory sense par excellence into the cyberworld of touch-screen digital devices), with increasing importance assigned to videogames, as they reach the age of three or five, taking up a lot of their time, enough for addiction. Certainly, complete familiarity with the ubiquity of electronic screens, big and small, domestic or other, in watches, toys, appliances of all kinds, every conceivable technological object, 24/7, can be achieved easily. In fact, from the first thing in the morning to the last thing at night we spend our lives looking at electronic screens, TVs, computers, tablets, smartphones, and so forth, but we are also surrounded by them in offices, shops, cafes, buses, cars, planes, and so forth, the number of places that can remain e-screen free diminishing all the time. Digital toys of many different varieties are supplanting long-established analogue ones, and electronic screens and mobile devices have become ubiquitous in domestic and extra-domestic settings for children from a very early age. It has been claimed that thereby children's relationship to the physical world (insofar as it is not yet transfigured digitally and virtualized) may change, and this is a topic that deserves investigation in cognitive–developmental psychology. And what will be the effect of giving a four-year old a Kindle so that he becomes familiar with e-books well before becoming acquainted with printed books?

A few years ago, a minister of education in a EU country pronounced on an official occasion: 'the school is a firm'. By the same token, one could also say, 'the university is a firm' (not just private universities), 'the Roman Catholic Church is a firm' (not meaning just the Vatican Bank), the 'state is a firm' (the state as such, generically, not just klepto-cratic or patrimonialist states), or even 'the family is a firm' (not just 'family firms'), even though conventional textbook economics in the West (e.g., Paul Samuelson's textbook) used to distinguish sharply between 'firms' and 'households'.

Without necessarily making categorical assertions of this kind, all these institutions, schools, states, churches, families (and practically all other social institutions) have been increasingly envisaged as instances of, or as embedded in, markets, market-oriented organizations or simili-markets. In part, certainly at the ideational level, this is because of the impact of modes of analysis – or at any rate legitimated by the ex post modes, which have become increasingly influential in the social sciences and, prescriptively, in public policy, especially in the last four decades or so (though a number of the canonical works deemed as the fountainheads of these modes of analysis date even further back). These have included the economic theory of democracy, the economic theory of the family, the economic theory of marriage, the economic theory of public administration, or public choice theory, the economic theory of law (the Law and Economics movement), the economic theory of crime, the economic theory of terrorism, the economic theory of nationalism, the economic theory of language, the economic theory of the university, the economic theory of science, the economic theory of doctor–patient relations, the rational-choice theory of religion (and indeed rational-choice theories of every kind of social institution and political practice) and so on.

All kinds of human interactions and patterns of behaviour have been subject to these modes of analysis, which have invaded the terrain once occupied by social sciences other than economics and by and large have superseded or subordinated non-market types of conceptualization, at least in the most highly ranked journals of the social sciences. Still, the most important consideration here is their influence, direct or indirect, proximate or remote, mediated by innumerable and prolific think tanks, especially in the United States and the UK, and even in France, on thinking about public policy.

The old language in which one spoke in, at least occasionally, reverential terms of the mission, functions, role, or vocation of, say, universities or schools or education in general, is replaced by, or at least subordinated to, language in which the mundane exigencies of markets are ascribed priority of emphasis and, if one says that the school (or another sort of public educational institution) is a firm, unqualifiedly, one may well be implying that the market is paramount in defining what can or should be done in the life of the institution, not the moral economy of its traditional understandings. In effect, it means the de-legitimation of the language of public purpose, of a public service ethic, of respect for intrinsic values, for the market is of course the prime site of extrinsic values, of instrumentality, of neutralization. It may be remarked that neutralization, or 'adiaphorization' in earlier parlance, in the wake of the wars or near-wars of religion had rent Europe in the sixteenth century of non-economic considerations, interdictions and constraints. It is also noteworthy that, in the West over the last centuries, neutralization was one of the sources of the expansion of the market sphere.

The use of the verb *to be* in such official assertions merits some reflection. The wise Adam Smith who, besides his work in economics and morals, wrote on epistemological and other philosophical topics, once stated that the verb *to be* was 'the most metaphysical of all verbs'. Indeed, many thinkers have sought strenuously to avoid this verb as particularly conducive to hypostatizations and other linguistic delusions or 'bewitchments' of language, detracting from our capacity to think critically, not least in the most polemical or emotionally fraught issues arising in public affairs.

An American movement of thought in the twentieth century, more recently calling itself General Semantics, went further and sought to prohibit and even eliminate this verb altogether from their work, regarding it as the logico-grammatical matrix of all essentialisms, with all the pernicious sociopolitical and even civilizational consequences that may stem from such prepossessions. They even constructed a language without this verb, at least explicitly, a Being-free language, so to speak, for indeed the verb *to be* is not only the most metaphysical of all verbs, but it is also the case that the bulk of Western metaphysics, especially ontology, produced in Indo-European languages, could not prescind from it. Be that as it may, the use of the verb in an official context in the way I cited seeks to convey even more than a 'persuasive definition' for, when spoken with the authority of a ministerial office, it becomes an injunctive and exclusionary term, even a performative one.

Church and State

For coherent market libertarians, churches (or religious confessions in general) ought to compete in the free market without legal restrictions on entry – such as those imposed by some liberal states, let alone de facto or de jure theocracies[27] – concerning authorization, registration, fiscal concessions and public law guarantees comparable to those that historic confessions have enjoyed, even if they have been losing privileges through disestablishment or by other means. For a libertarian who sees the free market as the paradigm for all worthwhile modes of social activity, there is no reason, in principle, so long as the criminal law is not infringed upon (and perhaps this body of law should be minimalist, as befits a 'minimal state'), why it should not be allowed to any individual to register a confessional enterprise or enlist himself/herself as a minister of whatever religion (not necessarily monotheist or deist, to be sure, paganism or Buddhism being accepted widely in the UK), selling or buying belief statements, creeds, breviaries, liturgies, ceremonies, hymns, and so forth, registered or even patented as intellectual property. For example, policemen in the UK who profess paganism (now a recognized religion) are granted leave to celebrate pagan sacred occasions, on a par with the leave for religious holidays of Christian confessions. Such religious associations can have a purely virtual existence, such as the 'radio churches' which flourished in the United States and Brazil in the thirties, and today 'online religion', not just televangelist, still has a lot of scope for growth. Conversely, any person, any citizen, should be at liberty to opt for any of the claimants in the religious market, possibly choosing the most promising offer of salvation (as has already been described by eminent sociologists of religion), choose other options afterwards, even more than once, trying them all out perhaps, demand new options, or simply not choose any of the 'products' in the multi-religious market. Needless to say, there is now truly a market, not just by conceptual analogy, with lucrative religious enterprises selling therapies and counselling, consolation and hope, and other 'products', behaving just like any private corporation, except that, on the whole, one can join and leave more freely. 'Pastorpreneurs' continually seek to shape better offers in the highly competitive religious market. Alternatively, any person could opt for one or another of the 'products' available in the irreligious or non-religious market which will surely expand in the next decades (atheisms – given the quarrels amongst atheist groups in recent years, as in 'Atheists against Dawkins', the plural is fitting, anti-theisms, anatheisms, 'Christian atheisms', agnosticisms, scepticisms, solipsisms, 'ethical culture', 'sceptical religion', etc.). This path would lead to the most complete expression of the free competitive dynamic market that America embodied with its changing religious landscape of sects and denominations. Historically, of course, non-exclusivist religions like classical Hinduism, Confucianism, Shintoism and so forth, allowed practitioners to follow several religious ways concurrently. Of course, there have long been Catholic atheists or Catholic pagans, venerating the church as an institution, defending it to the utmost, but not subscribing, at least inwardly, to its belief system, and sometimes rejecting it openly – as the example of Charles Maurras, such an influential figure over almost a century, shows. In the past they have been associated with 'counter-revolutionary', nationalist or para-fascist politics in France, but this need not be so, for there has been some talk of Catholic atheists of a new sort in the last decade. These seem to be intellectuals of Catholic background who have been impressed by Pope Benedict's philosophical discussions with philosophers such as Habermas. Many people claim that the Pope won on points. There have been also atheist Anglicans, the noblest one being the philosopher McTaggart, who was no materialist. It is also noteworthy here that religious affiliation has long ceased to be registered in the passports of Western citizens. Still, it was only a few years ago that Orthodox religion ceased to be inscribed in all Greek passports, as if religion and nationality were coextensive ascriptive attributes of Greek citizens. The recrudescence of ethno-religious nationalism in recent years may bring this kind of practice back in some countries. Recently, an Israeli writer succeeded finally in getting registered as a Jew in the ethnic sense but without belonging to the Jewish religious community; before this case, all such attempts had failed in the courts of the land.

Turning to the state, if the market libertarians have their way, we will pay direct taxes, local or national, if at all, to the state as a firm – a state, at any rate, that is run as much as possible according to the procedures and protocols of commercial companies, and whose classical activities have been extensively hived off to private or hybrid companies, hollowed out in order to provide greater scope for private entrepreneurs. Even before these trends came to the fore, the behaviour of civil servants, political parties and voters had been analysed by influential political scientists in the rational-choice paradigm in standard microeconomic terms, by analogy with the behaviour of entrepreneurs, firm, and consumers, that is, as actors in simili-markets, even if money was not actually circulated, as in kleptocracies, and literal vote bribery.[28] The analogy had already been put forward by Joseph Schumpeter in 1942, and in similar ways by other Austrian economists of sharply different orientations, in the case of free competitive parliamentary or presidential elections, and was worked out in systematic terms little more than two decades later by Anthony Downs and others, which many political scientists and economists have found persuasive. Our relationship with the state, if indeed the word itself survives, will be in the nature of a venerable relic. If current trends prevail, it will be much more like that of consumers/customers in connexion with the provision of certain classes of desired services by a firm, even if a somewhat *sui generis* one,[29] or a consortium of firms, as the unity of the state/government is sundered, though not as the doctrines of democratic political pluralism envisaged. Whereas the very idea of citizenship normally entailed an exclusive bond with a territorial state, even if not necessarily also individuated in terms of language or ethnicity, the *quondam*[30] 'citizens' morphed into consumers/customers, or 'citizen-consumers', as it has been called, in an astonishing move away from classical conceptions of citizenship, for certain 'packages' of services, which may be supplied by non-territorially defined state-less agencies: those consumers/customers may not feel bound in the same way to a given territory, even if they have lived there all their lives. The example of the wealthy is there, moving to the countries where taxes are lowest, if not to 'tax havens' and, indeed, the more ideologically motivated seem tempted by the ventures of 'seasteading', floating tax havens, whereby they may dwell outside territorial waters as long as possible, and thus be free from the legal and fiscal constraints of even the most market-cherishing or market-fearing states, not owing any loyalty to any nonmarket entities. In an ideal world, from the standpoint of market libertarians, all adults should be able to choose freely in the international market – independently of their past – the best state offer of the day (insofar as states still exist, perhaps as 'minimal states'), just as great companies and entrepreneurs do, even now, with respect to the best 'fiscal jurisdictions', fleeing from the countries with higher direct taxes on income and wealth, attributed primarily to social expenditures (even if defence may in fact be a very substantial claimant on state budgets). This mobility from state to state of consumers/customers, especially taxpayers in the higher income brackets, analogous to the mobility of companies seeking the least exigent 'fiscal jurisdiction', would be facilitated if the practice of registering yearly as wanting or not wanting to keep the nationality ascribed to them (possibly since birth) were to be institutionalized (as has been suggested by an eminent French sociologist, amongst others).[31] Endless eulogies of the 'Republic of Choice' seem quite unwarranted unless the 'choice of republics' were to be made as free as the choice of consumer goods, or brands, or something like it, since the choice such writers commended is based on the template of consumer choice. Market libertarians prescribe a world without any binding ties to others (though they owe stringent and lifelong duties to oneself, as in the concept of Absolute Selfishness put forward by Ayn Rand, the philosophical mentor of so many top American financiers)[32] except those of free but strictly temporary contracts with others, contingent on mutual consent (though not all contracts are like marriage today, at least in some Western countries, revocable on demand by one of the two parties, with minor restrictions as to time elapsed, thus going beyond mutual consent).

States (or unions of states, like the EU) have been generally jealous of their monetary sovereignty. But market libertarians have suggested that private enterprises could issue their own currencies, which would compete in the same territory (a national territory or that of a multinational federation). The denationalization, or de-statization or privatization of currencies, was advocated at length by Friedrich Hayek and other economists of lesser fame. It is perhaps one of the least insistently voiced recommendations of the Hayekians and like-minded market liberals or libertarians (although it has played a role in their campaigns against the European Monetary Union). Even today, the agencies that predominantly 'produce' money, the monetary aggregates that appear in national statistics, are predominantly the nonstate banks, or were before so many banks were temporarily nationalized and 'quantitative easing' became a frequent practice.

One final point about states is worth making. Some time ago, it was reported that British citizens who had applied for their passports, received an apology for the delay in their issue, addressed to 'Dear customer'. It is true that payment for passports, even legal ones, has become routine, but in the past this fact did not imply in anyone's mind that the exchange of money for a passport to which one is legally entitled, as a citizen, as proof of nationality/statehood, was simply a commercial transaction like any other, a cash nexus without any serious non-monetary import. It involves a citizenship bond, not a market relation. Possibly birth certificates, death certificates, state pension statements, juror notices, messages to electors and so forth, are already sent to 'Dear customer(s)', or soon will be. One wonders whether the Ministry of Defence will refer to dead soldiers as 'our late customers', unless we deem the state as no more and no less than a firm like any other, UK PLC, and so on, though demanding at times a special kind of brand loyalty, or corporate loyalty, and charging a high 'transfer price'. In a world even more extensively structured by market libertarianism than now, passports might be replaced by laissez passer certificates, enabling us to travel beyond former national territories, issued by commercial firms, perhaps contractually tied to transnational security companies, answerable to no state in particular, to protect us in the course of our long-distance travels.

These practices belong to a widespread trend in the last decade or so, to replace any number of hallowed terms like patient (in hospitals or clinics), resident, passenger or traveller (in planes, ships, trains or buses), library reader, museum visitor, or client, often pertaining to fiduciary relationships with liberal professionals (lawyers, solicitors, accountants), not to mention 'citizen', uniformly and universally replaced by 'customer', and thus being so addressed verbally and in other kinds of communication (terms like student, pupil, trainee, educand, parishioner, sports spectator, cinema patron, and so forth, are under similar threat). For example, the new national curriculum in the UK includes, amongst its 'basics', besides history, science, geography, and so forth, 'citizenship'. What does it purport to impart? '[Children] should be equipped with the financial skills to enable them to manage their money on a day-to-day basis as well as plan for their future financial needs' (*The Independent*, 8 February 2013, p. 5). That is, topics addressed in 'home economics' courses in other countries. There is nothing about civic liberties and duties, parliamentary democracy, voting, civic participation, the place of the national state in a world with multiple transnational obligations, and so forth. Citizenship has become equated with financial prudence in one's own affairs. Is it supposed that this ability could be transferred to the management of national finances just like that?

These trends have crept in incrementally, but this has been a practice mandated by management or government directives and fostered by a climate of state-encouraged market-speak. A very recent comment by a British surgeon illustrates how a professional imbued with the old service ethic felt about the new linguistic practice: 'In the 1980s, the chairman of our hospital's management committee rebuked me for using the word "patient". He said I should call them "customers". I had to explain that a customer is someone who can go elsewhere if he does not like the service; a patient is someone to whom the doctor owes a duty of care' (Barton 2013).

The transformation of the language of forms of address or standard relationship-defining terms, with the prohibition of old terms and their replacement by new, state-approved ones, has been most of the time the prerogative of radical regimes of the Left or Right (the mandatory 'citoyens' in Revolutionary France as the unique form of address for everyone was perhaps the first such example, but many others were added in the twentieth century, often involving personal pronouns, especially the second person where a 'T/V semantic' (tu/vous}, the semantic of the 'pronouns of power and solidarity', prevails). It is striking that we find here in a market democracy an attempt to enforce a uniform sense of customership throughout a diverse world of health services, education, transport and so on, as constant reminders of the cash nexus as the underlying reality of a lot of, if not all, social and political connexions. In a way, we encounter here a reform, a market reform, albeit unannounced, of language itself, at any rate of standard linguistic practices, to enforce a kind of pedagogy of marketness (though 'customer resistance' has made some companies backtrack on the linguistic practice, and we are now often called 'passengers' when travelling in commercial airlines or in private company trains and no longer 'customer'). The sociological import of these trends will not be lost on those familiar with Herbert Spencer's concept of 'ceremonial institutions', the standardized linguistic markers of respect, parity, deference, recognition of equality or hierarchy, relationship or status definition (Franzen 2011). To be sure, social movements can be responsible for pervasive changes in key denominations. Feminism's introduction of the title *Ms*, is an important example. The diffuse trends into so-called political correctness were partly fostered by anti-discrimination, anti-stigmatization sentiments attached initially to anti-sexism and anti-racism, but generalized to protests against ageism, ableism, weightism, heightism, heterosexism, and so forth, with varied outcomes and much ribaldry, but with some successes in diminishing the habitual, unthinking use of offensive terms and replacing them with more neutral, freshly minted ones. Under the Blair administrations the promotion of certain linguistic usages was much noted and was the subject of scholarly studies (Fairclough 2000). But, besides politically induced changes, Information Technology Companies and especially social media networks have also played a role in the 'corruption of language': 'the transformation, courtesy of Facebook, of the verb *to like* from a state of mind to an action that you perform with your computer mouse, from a feeling to an assertion of consumer choice, is an example of this'. And liking, in general, is commercial culture's substitute for loving. The striking thing about all consumer products – and none more so than electronic devices and applications – is that they are designed to be immensely likable. This is, in fact, the definition of a consumer product, in contrast to the product that is simply itself and whose makers are not fixated on your liking it. I am thinking here of jet engines, laboratory equipment, serious art and literature. But if you consider this in human terms, and you imagine a person defined by a desperation to be liked, what do you see? You see a person without integrity, without a centre. In more pathological cases, you see a narcissist – a person who cannot tolerate the tarnishing of his or her self-image, which not being liked represents, and who therefore either withdraws from human contact or goes to extreme, integrity-sacrificing lengths to be likable.

Data and Information

Recently, there was a case in the Supreme Court of the United States that concerns whether physicians should or should not be legally obliged to sell to pharmaceutical enterprises their clinical notes relating to their prescriptions for patients, filed as digital data, should the pharmaceutical enterprises so request. As in all cases where it is required that data be stored in digital format, a practically universal requirement in Western countries, it is to be expected that, eventually, the data will be accessed by third parties, malicious or otherwise. Most often this will take place without the knowledge of the persons whose data were thus filed, and even without the knowledge or consent of the professionals who committed themselves in good faith initially to ensuring their privacy and safety. The access to such e-data will take place by legal, semi-legal or illegal means, at least through hacking, data-mining or cybertheft. Access to data for commercial effects is sometimes guaranteed by law, but obviously there is considerable demand for an extension of this access. In the case of data protected on the grounds of national security, cybertheft, hacking, cyberspying and cyberwar become ever more sophisticated. Indeed, one might argue that we already live in an age of permanent cyberwar, at any rate of continuous cyberespionage, and never-ending attempts to breach the cybersecurity of foreign state agencies to access secret data of various kinds, sometimes with weapons such as potent computer viruses or worms disabling physical installations like nuclear reactors or centrifuges. Of course, there occur also cyberattacks on commercial firms, and recently on the *New York Times* and the *Wall Street Journal*. Presumably all organizations that have 'dematerialized', that is, digitalized their data, files, archives and communications, dispensing with 'paper trails', are potentially susceptible to cyberattacks. Be that as it may, the extraction and storage of digital data makes the commercial and criminal exploitation of personal data easier, not only in the case of clinical data, but throughout our cyberlives, encompassing today practically all our commercial transactions in 'advanced' countries, our e-mails, SMSs, online chats, Internet visits, not just Facebook and other social media networks, photos online with Instagram and the like. All episodes of electronic communications, and a very large proportion of Western populaces that have become a sort of 'e-demos', which does not ensure the coming of a 'digital democracy', commit, so to speak, digital acts of one sort or another by the hundreds or thousands per day (e-mails, SMS, Facebook visits, phone calls on mobiles, etc.), cumulatively building up considerable 'digital footprints' and 'digital shadows' of their authors.

Of course, enterprises like Google and other major firms in information and communications technology (ICT) leave a considerable carbon footprint, much as they seek to reduce it. Thus, the energetic costs of ICTs cannot be simply ignored, nor information taken in a somewhat ethereal fashion. Even the great enthusiasts for the Data Revolution (latterly the Big Data Revolution) enabled by the huge 24/7 flows of messages of electronic communications in commercial and professional activities, not to mention personal matters, and of course machine-to-machine communication (M2MC), which in volume already exceeds human-to-human communication and will grow with the development of the 'Internet of things', recognize that all digital information is hackable, theft-prone, and commercially exploitable, though this is not to say that it cannot be put to positive, if unintentional uses, by third parties for the common good, such as the detection of earthquakes, epidemics and other natural calamities via detected traffic intensities. One cannot fail to mention the use of such data by and for the 'surveillance states' already in being, whose appetite for the tapping of digital communications appears to be unlimited and requires the building of extraordinarily large structures and complexes, while the improvement, even the perfection, of encryption technology remains a high priority. Unbreakable codes, possibly post-digital, of quantum information, as quantum computing is advancing, are the Holy Grail of the digital age. All digital information produced and circulated in the human–machine world sooner or later enters the circuits of commodification, licit or illicit, in some respect or other. In any case, the end of personal privacy as a result of the Digital Revolution has been proclaimed repeatedly by the *digerati* over the last two decades or so and emphatically reasserted by commentators regarding the recent explosion of smartphones, now outpacing unembedded computers. Of course, notable students of the subject have argued for some time that we have to stop our lamentations or forego our misgivings on this score, and simply accept and make the best of, the emerging 'transparent society'. Transparency, or total disclosure of everything, or at any rate of everything digitally committed or leaving digital traces, to potentially everyone will be the default.

Another related trend is that electronic communication, with the Digital Revolution, has become a great multiplier of and accelerant for universal commodification. Universal in geographical extent and, even more importantly, in the ever-increasing scope of what can be included in the process of digitalization and electronic/photonic transmission: messages, visual images, sounds, ideas, memes, news, sex, physical goods or scripts for their making, formulae, objects, properties, geospatial information, weapons and so forth. Financial globalization, a key lever for globalization in the wider sense from the 1970s in particular, even though we still live in a 'semi-globalized' world, is quite a bit less 'flat' than has been proclaimed and, in turn, has become a mechanism for accelerated commodification. Financial instruments, ever more abstruse and even incomprehensible to their own users (incomprehensible, but carrying the expectation of large gains), operate in stock exchanges and are processed in ever-greater proportion, as 'high-frequency computer trading' (also called 'algorithmic trading' or 'black-box trading'), that is, algorithms devised by mathematicians and physicists with PhDs from the best universities, the 'quants', who flooded into the financial services industry from the 1970s onwards (though the quants are but a sub-species of the *numerati* or 'super-crunchers' in all sorts of service industries and defence/security agencies). The proportions of the total-share transactions in such places processed by algorithms and not by human traders has increased to the point that, in the United States in 2011, they comprised in some instances something like 75 per cent of all transactions (Bowley 2011). Other stock exchanges have been urged to increase the proportion to 80 or 90 per cent, reducing the role of human traders to that of a residual category in this business, handling a small proportion of the volume and value of the financial transactions in such places. Taking into account the stress of these activities for human agents – some claimed 'to nearly die every day' to researchers, though actual deaths in any numbers have not been attested – perhaps this replacement of humans by computer algorithms, meatware by software, should be seen in a positive light, up to a point. However, the advantages claimed for fast, efficient processing of share deals by software programs have to be weighed against the financial and macroeconomic instability and the Great Recession starting in 2007–8 – for which the financial services industry is surely quite directly responsible, if unrepentant – preceded and accompanied by the Great Regression in the attacks on the progressive income tax and the trends in income and wealth inequality.

The 'robot traders' represent a kind of advance guard of the 'smart' robots, in a greater proportion than in most other major industries, and which have been assigned manifold tasks in the 'knowledge economy', a.k.a. 'knowledge capitalism' or 'digital capitalism', replacing the 'wetware' of human professionals, whose brains, are as yet innocent of neurotechnology. This is a field of intervention likely to develop in the coming years, though talk of 'neurocapitalism' seems premature and unable to process fast enough the bitstreams of data arriving every minute, every second, every nanosecond, with the maximum compression of cyberspace and cyber time. 'Robot scientists', as they were called when first designed by a team of geneticists a few years ago, have been making progress in the field in which they were initially assigned, that of molecular genetics, actually making new discoveries and not only corroborating received knowledge; they may take over much of the more time-consuming labours of graduate research assistants. 'Robot journalists' only started in 2011, and may be regarded as still only incipient, but seem very promising, especially in the areas of sports journalism, financial journalism and property (real estate) journalism. In any case, 'computer-assisted journalism', or 'computational journalism' are phrases that enjoy a certain vogue in the United States. A successful company in this endeavour, *Narrative Science*, boasts of an artificial intelligence platform that can transform any body of data into comprehensible newspaper articles (Fassler 2012). The interactive 'conversational robot' teaches someone how to speak English online. It is not clear to me which sort, since there is no phonetic World English, strictly speaking, though perhaps increasingly American accents will predominate. Algorithms in automata will undoubtedly be ever more widespread, replacing or complementing wetware in agriculture, livestock production, industry, commerce, transport, the workplace, surgery, science, warfare and outer space exploration.

And of course there is hardly any significant industrial, commercial, financial, or intellectual activity in the sciences, humanities (the 'digital humanities' movement has made considerable strides, so far mostly in the United States) or even in the arts (above all in architecture) which is not in some fashion computer-assisted, some indeed computer-assisted through every phase, even if only some of these activities are designated by acronyms such as CAD (computer-assisted design) or CAM (computer-assisted manufacturing) to mark their sharp difference from the common garden varieties of such endeavours. Before too long, such acronyms may not be needed any longer, as they will be the default modes of design and manufacturing. We have also CAPI (computer-assisted personal interview) in organizations, computer-assisted proofs in mathematics, computer-assisted 'discovery' in the American legal process, and soon many, many more CAXXs, in every field of social practice. It is noteworthy here that the expression 'computer-assisted' is standard. One finds fewer occurrences of the expression 'computer-controlled' (as in 'numerically controlled manufacturing'). Perhaps this is because of a generalized euphemistic tendency to avoid reminding ourselves of how far we have gone along the road of granting computer programs not just monitory, but more than that, regulatory, and even practically prescriptive roles.

Armies and Police

If there are still armed forces bound to states, they will be more and more like bands of mercenaries, or volunteers for the armed forces, as compulsory military service in peacetime has been abandoned in most Western countries since the 1970s. The international market for weapons, still flourishing today, especially but not exclusively the more advanced ones, which has been so important in the world economy for decades, during and after the Cold War, may be supplemented by an international market for military, or 'specialists of violence'. Already tens if not hundreds of thousands of armed civilian security personnel work in dangerous areas in regions like Afghanistan or Iraq, and elsewhere, although they are not generally called mercenaries.

A shortage of suitable volunteers – even though criteria of suitability, such as those pertaining to age, height and weight – have been loosened quite a bit in recent years, may be compensated for, as always, by qualitative improvements in fighting capacity (as well as by the maximization of firepower). Thus, some current Pentagon programmes aim at the making of 'super-soldiers' ('soldiers 2.0'?). They would be provided, for example, with 'exoskeletons', with their own energy source, enabling them to carry with ease far greater loads than hitherto, and at a speed comparable to that ensured by the seven-league boots of classical fairy tales. In addition, it is envisaged that a wide array of physical and cognitive enhancements will be afforded by genetic, surgical, neurological and pharmacological means, as well as by coupling them with varied apparatuses. Latterly the case for moral enhancements by similar means has also been made. Such (nongenetically) modified humans would be formidable warriors, though presumably many of these enhancements would be transferable to civilian life in due course, and perhaps there will be a race to provide commercially for those who want to become enhanced humans in this fashion. Drugs, for example, will ensure a state of high alertness and cognitive acuity for many, many hours, if not days. Indeed, some drugs in fact have been compulsory, on pain of dismissal, for pilots in the American Air Force, for certain kinds of flights. Such drugs are not yet compulsory in civilian work life. These will be complemented by a plethora of prosthetic electronic devices to enhance vision (telescopic, microscopic, and all-around, 360 degrees) and hearing, endowing soldiers with a super-sensorium in addition to their super-motorium. Of course, these electronic super-senses are likely to be deployed for civilians sooner or later: Google glasses, and smartphones with microscopic vision, and other optical instruments with both microscopic and telescopic capabilities of high resolution, when they arrive, will pioneer such devices in civilian life, which might in due course become the standard equipment of the ordinary citizen.

In addition, there are armed robots, unmanned tanks, unmanned helicopters, planes or drones of widely varying sizes, from that of jetliners to tiny, almost invisible ones, weighing as little as 16 grams, carrying missiles and other weapons as well as super-cameras. They are assisted by swarms of cyber insects for reconnaissance, even unmanned naval craft (presumably also in a variety of sizes), surface and underwater vessels, amphibious, and so forth. Drones, or unmanned aerial vehicles, are of course already widely deployed, supposedly for precision killing or assassination from the air or minute surveillance over a number of theatres of war, though there has been much consideration of the ethics, legality and even the constitutionality, of their deployment by police and security agencies in domestic jurisdictions. For example, British troops in Afghanistan have been supplied with drones, a special kind, Black Hornet Nanos, which is palm-sized, hand-held, and can go around corners and obstacles to detect hidden dangers.

There has already been quite a lot of discussion on the ethics of a robotized war, in which robotic weapons platforms fight it out amongst themselves, and make decisions not only to destroy other robots, but also to use their firepower to kill humans, bomb and wipe out human populations. Decisions will have to be taken *in loco* by smart robots,[33] thus contradicting Isaac Asimov's 1942 'laws of robotics' as well the much older 'laws of war'. In any case, robotized weapons are likely to play an increasing role, even if in short of wars between robots and robots exclusively if, in the last analysis, some human commanders still control the encounters. The ethics of autonomous robots in warfare has naturally been the subject of extended discussions (Cartwright 2010). It has emerged as a particularly urgent branch of the recently denominated field of Artificial Ethics, the ethics of artificial entities and 'agents', though they might be called more properly 'quasi-agents', or 'actants', 'smart', and getting smarter all the time, sometimes accredited with autonomy (intelligent cyber weapons, too, would fall within the purview of this kind of inquiry). Needless to say, a bonus is, surely, that weaponized robots do not yet commit suicide after they have been in action (though, of course, they can be programmed to do so), unlike human soldiers. American soldiers on active duty today die more from suicides than from combat in Afghanistan (one suicide a day on average), even though there are counselling programs aimed at suicide prevention, set up years ago. Nor do robots suffer from PTSD, though doubtless they could be programmed or reprogrammed (by their makers or possibly by cyber saboteurs) to undergo something like it, for real or as fakery.

Meanwhile, as we noted, electronic cyberwarfare, offensive or defensive, has been taking place in the last few decades, conducted by a certain number of countries (the United State, China, UK, Russia, Israel, among others), between hackers and super-hackers, between different sets of computer malware (viruses, worms, trojans) accompanied by cyberespionage and cyberterrorism and, of course cyberdefence and cyber counter-terrorism, in their industrial, commercial and political variants. Not just state-to-state, but state-to-firms, for the purposes of industrial espionage. This is an age of permanent cyberwar, at least of a 'cold' version of cyberwar, so to speak. As yet, without too many episodes of cyber physical attacks, such as disabling nuclear centrifuges – as has already occurred, via the computer virus denominated Stuxnet by a cybersecurity firm, a name that stuck, a harbinger of even greater things to come – there is possibly a turning point in our cyberworld. Some scholars have identified this Stuxnet event and what it portends as a turning point in the history of our computational–informational industrial and scientific world. The executive editor of the journal *Scientific American*, for example, called the advent of Stuxnet a turning point in the history of technological civilization.[34]

Conclusions

In this brief narrative, I have had necessarily to omit quite a number of extremely important areas such as media, sports, the workplace, economic organizations, the legal system, public administration, nutrition and tourism, amongst others. But even with these very large omissions, the topics addressed may provide reasonably illuminating mini samples of the saturation of all spheres of human life by markets, privatizing and invading previous commons, under the pretext or with the aid of allegedly unstoppable and accelerating technological innovation now often termed 'disruptive', but in a surprisingly commendatory fashion.[35]

Notes

1 John Burdon Sanderson Haldane (1892–1964) was a scientist of British descent who was later naturalized as an Indian citizen. His book, *Daedalus*, which was initially read as a lecture in 1923, speculated about the future of human evolution through ectogenesis (Haldane 1924).

2 On the 'baby business' as a techno–mercantile complex with varied sub-markets, see Debora L. Spar's *The Baby Business: How Money, Science, and Politics Drive the Commerce of Conception* (2006).

3 On the persistence of eugenics, of eugenic doctrines certainly, over more than a century, not least among scientists, and especially its recrudescence in the West in recent decades as techno-eugenics or market-eugenics, see chap. 9 of my book *Experimentum Humanum* (2011).

4 This is the thesis of the 'bioethics of birth' advocated by M. Rothblatt in *Unzipped Genes* (1997).

5 Mécary's statement roughly translates to the following: 'To allow women, who consent in a legal framework, to offer a child to a couple that cannot have one, would it not be the greatest feminist subversion that one can imagine: to finally free oneself from the duty of being a mother?'

6 *Jus sanguinis* is a Latin phrase that means 'right of blood'.

7 *Raison d'être* is a French phrase that means 'reason for being'.

8 Sir Ian Wilmut (b. 1944) is a British embryologist who led the 1996 Dolly cloning project, wherein a female domestic sheep was cloned through nuclear transfer.

9 *Corpus mysticum* is a Latin phrase that means 'mystical body'.

10 Peter Singer (b. 1946) is an Australian moral philosopher known for his work in bioethics and his utilitarian stance on animals.

11 Lady Mary Warnock (b. 1924) is an English philosopher known for her work on reproductive rights, existentialism and ethics. Warnock served as chair on the UK Committee of Inquiry into Human Fertilization and Embryology, which resulted in a 1984 report, referred to as the Warnock Report, that led to key regulations in infertility treatments, such as IVF, and embryology. (James Stacey Taylor, *Death, Posthumous Harm, and Bioethic* (New York: Routledge, 2012), 228 pp.)

12 See my account of the history of eugenics in chapter 9 of my book *Experimentum Humanum.*

13 On experimentation with human subjects and the eventual coming of this master precept, see chap. 6 of my book *Experimentum Humanum.*

14 Prima facie is a Latin phrase that means 'at first appearance'.

15 See the references and links in the striking article by Bill Gleason 'Do People Have a Moral Obligation to be Research Subjects?' (2011).

16 *Homo senectus* and *Homo economicus* are Latin phrases that mean 'old man' and 'economic man', respectively.

17 Matthijs Pontier gave a talk in February 2013 with the arresting title 'Dating a Robot is like Dating a Man: A New Perspective on the Turing Test'.

18 Claude Bernard (1813–1878) was a French physiologist who coined the term 'milieu intérieur' to refer to the internal fluids that protect the tissues and cells of the body. David Émile Durkheim (1858–1917) was a French sociologist known as one of the founders of the field of sociology.

19 *Pons asinorum* is a Latin phrase that means 'bridge of asses', referring to a geometric theorem and, more broadly, a problem that represents a threshold in the testing of one's abilities.

20 *Memento mori* is a Latin phrase that indicates a reminder of death, literally translating to 'remember that you must die'.

21 By analogy with the French terms 'bibliothéque' or 'discothéque' and their counterparts in Spanish or Portuguese. It would be easier to coin an appropriate term in a Romance language and, in my original paper, I coined the neologism 'tanatoteca'.

22 *Pari passu* is a Latin phrase that means 'on equal footing'.

23 I thank Marcos Barbosa de Oliveira for a close reading (and a correction) of the original version of this text.

24 *Laïque* is a French term that means 'secular'.

25 *Desideratum* is a Latin term for 'something that is desired'.

26 Not everyone concurs with this pessimistic view.

27 De jure is a Latin phrase that means 'according to law'. It can be contrasted with the Latin phrase de facto, which refers to something that exists 'in fact', even if it is not prescribed by the law.

28 Although voting in parliamentary and presidential elections (if by popular suffrage) has been compulsory in some countries for a long time, most countries do not provide incentives, even minimal ones (like holding general elections on Sundays when most people could vote without loss of wages (before 24/7 work schedules), even though the proportion of electors casting their votes has been falling for several decades in several major as well as minor democracies.

29 *Sui generis* is a Latin phrase that means 'of its own kind'.

30 *Quondam* is a Latin term that refers to the past, meaning 'formerly' or 'at one time'.

31 Alain Touraine. I do not know if he still holds this view.

32 The influence of her writings did reach beyond the United States, even into unpropitious terrain, such as France, decades ago.

33 *In loco* is a Latin phrase that means 'in place'.

34 Mariette DiChristina has been the editor-in-chief of *Scientific American* since 2009 and is the first woman in the journal's history to hold this position.

35 And, it is often claimed, it should not be stopped, even if we could, and we cannot, and indeed, why should we? These phrases are often found in the writings of scholars or perhaps we should call them in terms of the role they play: publicists, public intellectuals or policy intellectuals: these are vulgarisms, but nevertheless asserted with the utmost emphasis, investing them with an aura of apodicticity.

Chapter Two

TECHNOLOGY SUBLIME: PATHS TO THE POST-HUMAN

It were no slight attainment could we merely fulfil what the nature of man implies.

Epictetus

Eris sicut dii.
We shall be like gods.
Ex machina libertas.

Transhumanist mottoes

Many current projects and visions about the future of humanity and the human being, allegedly well-grounded on the current moving frontiers of scientific and technological advance, envisage not only the substantial amelioration but also the transformation or surpassing of the present condition of the human species.[1] This process would be accomplished through the 'enhancement' (a key term of this discourse) of its genome, in conjunction with developments in computation, Artificial Intelligence and the technologies of information and communication (ICTs),[2] regarded as paramount, and often as the vehicles for the next and final stage in the post-biological self-transcendence of the human being, indeed, as they say with respect to the *singularity* in question, beyond which, as the name suggests, nothing can be known, even in principle. On the way, in the next few decades, we would acquire:

- a 'new body', version 2.0, to be followed, presumably, by versions 2.0 plus, 3.0 and so on (Kurzweil 2003);
- a 'new brain' ('the merger of biology and technology'); a 'new mind' ('the merger of psychology and technology');
- a wealth of 'smart genes' or super-alleles;
- 'super-senses' aka bionic senses (Geary 2002);
- superlative cognitive competences;

and secure 'radical life extension' but with unimpaired quality of life (Wade 2009). In other words, we would acquire the next best thing to biological immortality before attaining genuine, post-biological immortality as virtual beings, a kind of eternal life without life, in the organic sense of the term 'life', or indeed of the 'spiritual' (Broderick 1999; Kurzweil 1999; Tipler 1994).[3] To be sure, in this process we will also have to face up to the increasing cognitive abilities and demands for rights (moral, legal, civil, political) of nonhuman beings (robots, androids) we have been creating, endowed with artificial minds, artificial intelligence of ever higher grades, and share the cyberworld with them, or indeed merge with them ever more completely, the 'merger of flesh and machine' being an important theme in the arts and in the para-scientific literature with respect to AI and kindred matters, starting with 'techno-babies' or 'cyborg babies'; being a cyborg is, surely, a matter of degree as, indeed, being a 'techno-baby'.[4] As well as 'merger', the terms 'fusion' and 'symbiosis' are also widely deployed in these contexts: the 'breaking of boundaries' seems to be regarded as, *eo ipso*, beneficent, a *desideratum*,[5] indeed, a major accomplishment of contemporary technology and even a kind of ontological bonus. In the end, what seems to be desiderated is not so much the breaking of this or that boundary (ontological, or onto-ethical, as it were, though *meontological* would be more appropriate since it involves nihilations of one sort or another (Fackenheim 1961); and are justified on grounds of the greater human good. Examples include the diminution of the evils of exclusion or limitation, the boundaries between the sexes/genders, between the mechanical and the organic, between brain and machine, between flesh and machine, between the species, and so forth, and *the very condition of boundarylessness as such*.[6] Now this condition has always been a mystical aim, the pursuit of which now is not a spiritual exercise, but a technologically mediated one, and not by techniques of the body, but by sophisticated microelectronic technology.

Science fiction? These outcomes, or most of them, are generally stipulated not only as possible, but even likely, or quasi-inevitable, within a time horizon of a few decades from now, certainly within this century, extrapolating from current trends (Greenfield 2003). To speak of these scenarios as possible worlds, or even as worlds one would like to see actualized, is one thing, but to raise the stakes to the necessary, to shift from the optative to the imperative mood, from the alethic modality of the possible to that of the inevitable, is quite another matter.

Claims about 'necessity' are often made with respect to big planetary threats, such as the challenge of global climate change and other issues that threaten the survival of the species. In essence, the claim is that such threats cannot be addressed without resorting to a global or, perhaps better put, radical, human change. Changes in institutions and practices, economic, social, political, cultural, are not in themselves adequate; instead what is also crucial is a consideration of the very being of the human being, of what counts as a human being – which, it is claimed, is becoming technologically possible through highly sophisticated means, within a time frame of decades, an intra-generational time frame, and not just through crude negative or even positive eugenics over many generations, or through state agency, as was widely conceived from the 1890s to the 1950s (Martins 2008). 'Necessity' is also construed in another sense. It is our *duty*, some argue, to perfect ourselves to the utmost, which means in our time, as contrasted with the understanding of both the Greeks and Christianity. Their goal is not moral perfection, or self-improvement in learning and sensibility, but availing ourselves of all the current or developing technologies at our disposal to reshape ourselves (genetically, neurologically, prosthetically) or fairly soon to be at our disposal, distributively and collectively.[7] It is our duty also, secondarily, to seek to accelerate the production of technologies to carry out this *moral imperative* of shaping human evolution in order to maximize our capacities (construed essentially as our cognitive capacities) and thus our power to shape and reshape human and nonhuman nature (Campbell 1995). The duty or imperative falls upon individuals in the first instance, though presumably the state would have a consequent duty not to interfere with the technologically mediated self-perfecting of individuals, even though it would imply choosing the new generations. They might also argue that the state's obligation in this context would be one of 'hindering the hindrances'[8] to the self-perfecting of citizens, which would still fall short of endorsing a prescriptive perfectionist political philosophy, according to which the state should actively pursue substantive ('thick') conceptions of the good life and the good citizen in the political community. In any case, political perfectionists need not endorse, let alone *mandate*, genetic engineering as requisites of the good polity, the good society or the good citizen. The liberal state, defined by its stance of neutrality over rival conceptions of the good, would presumably stand by as the processes of human enhancement and transcendence were taking place, surely unevenly, amongst the citizenry. To be sure, in recent years the liberal stance of neutrality, which in any case no liberal state has ever practised very rigorously, does seem to have been mitigated by concerns to promote happiness, voiced by some government ministers in the UK, and doubtless encouraged by the trend amongst economists to work out an 'economics of happiness'. But that pursuit usually takes human beings as they are, in their genetically or neurologically unengineered state.

Sometimes even, irreligious, post-Christian atheists use not just the language of moral imperatives, of 'oughts', but even the language of 'sin'; it would be a sin, they say, not just a moral wrong, not to do our utmost to perfect ourselves with the technologies of genetic and potentiation at our disposal, now and in the proximate future. This process of perfecting ourselves would bring about 'better humans', 'redesigned humans', 'more-than-human humans', indeed a new more-than-human *species*, with higher biogenetic and brainware capabilities and, subsequently or concurrently, a transition to a post-biological, computational, virtual mode of existence, where software will replace our meatware *completely*, in the company of other virtual beings, who were virtual *ab initio*,[9] who never went through an organic phase (Miller and Wilsdon 2006; Stock 2002; Naam 2005; Silver 1997; Kurzweil 2003).

The visions I have articulated in this text thus far are actually entertained by some workers in or commentators on the possibilities of substantial or 'radical' (their term) human transformation arising out of such booming fields as genetic/genomic engineering, the still-incipient field of synthetic biology (often characterized as 'radical' or 'extreme' genetic engineering) and 'reprogenetics', the combination of genetic engineering with the new reproductive technologies. Among the emerging programs are those on human reproductive cloning, the 'strong' research programmes of Artificial Life (computer simulations of organisms aimed not only at a better understanding of vital processes in detail but of life itself through the creation of life *in silico*[10] for, the A-lifers claim, we can only really understand what we can make), artificial intelligence (or machine intelligence, though the design of biological computers gives a new twist to the hardware/software contrast, as yet unassimilated in this literature), and the cognitivist research programme in brain–computer matching. And, as the mind *is* the brain on the current orthodoxy, this is really the brain–computer matching programme. Moreover, there are 'immersion technologies' like Augmented Reality, Virtual Reality and software simulation, especially in its more *outré* manifestos,[11] robotics (whether in its standard or in its evolutionary versions), neuroscience and incipient neurotechnology, and nanotech or atomtech (though we could also call it *nanoscience*). Sometimes commentators draw also on the more dubious scientific or para-scientific research programme of memetics (though memetic engineering need not be precluded thereby, and false theories can inspire successful practices, as false premises can lead to true conclusions). They are all in the business of 'inventing the future' (as is so often, and aptly, put by the practitioners and the commentators), 'inventing' our own future, reinventing future generations, human nature and the human being, and enabling new modes of personhood.[12]

Key to these inventions are 'techno-facts', which are facts generated or registered only by technology or with sophisticated instrumentation, many of which had never occurred before in the history of the Earth, and perhaps were observed or produced for the first time in the laboratory. Techno-facts may be seen as somewhat like Durkheimian social facts, with their defining or criterial properties of externality and constraint but also, of course, as Durkheim added with respect to social facts, endowed with an extraordinary appeal and lure, though in the case of technology today its *esse* is *fieri*,[13] rather than a simply and exclusively instrumental anthropogenic world at our service and command. It is difficult not to entertain the hypothesis that, given the current commitments to technoscientific endeavours in these and related fields like nanoscience or nanotech/atomtech, the generic biopsychological condition of *Homo sapiens sapiens*, the current biological or biologically grounded unity, unicity and identity of humankind, de facto, certainly, and eventually de jure,[14] may not be sustainable for much longer, even if we think in the relatively short run, in terms of decades, rather than millennia or even centuries. Note that pre-biotechnological transspeciation took millions of years, but now the time horizon has contracted substantially, and indeed the key role of technology has always been one of acceleration or time compression of natural processes.

For such extraordinary ontological shifts, even centuries would have represented amazing time compression or, to put it another way, acceleration and the acceleration of acceleration of techno-economic changes, local or global, compared with the time scale of biological evolution, with speciation requiring a time scale of the order of magnitude of millions of years or more, though the ensemble of human-induced, specifically biomedical technology-induced phenomena of 'hyper-mutations', bringing about the rapid emergence of super-weeds and super-bugs, antibiotic multi-resistant viruses and bacteria (what economists would call the negative externalities of widespread antibiotic consumption), have been characterized by some biologists as an 'evolution explosion', anthropogenic, but unintended and unwanted (Palumbi 2001).[15] This is quite apart from the continuous selective pressures exerted by human beings, the imperial species, on other species, most effectively through the ongoing Sixth Extinction in the history of life on Earth, which is to a considerable extent anthropogenic, directly or indirectly, a claim that does not appear strongly contested (independently of any considerations as to whether, or how far, global climate change is in fact anthropogenic and, if so, remediable to some degree; this might be owing to nonlinearities in the global climate system, which may imply that even a slight increase in human impact may have a wholly disproportionate effect, owing to positive feedback loops in this 'chaotic' system).[16] Or, indeed, compared with cultural change in previous epochs, a considerable number of time-series demonstrate a variety of exponentials, even at a planetary level, in terms of population and in a variety of indicators with doubling ratios of a year or two in Moore's Law of the doubling of computer power every 18 months, and its putative counterparts in a variety of fields. Moore's Law is not, of course a law, in the strict sense, but computational advances in the last forty years or so have occurred much as if it applied. So much so, that it can be argued that we may sum up recent and ongoing trends in technological change not just in terms of a law or laws of increasing returns and falling costs per unit of production, falling to practically zero marginal cost (laws to which economists have been paying more attention than ever, given their prevalence in high-technology industries), but even in terms of a law of not just increasing or exponentially growing returns, but a super-exponential Law of Accelerating Returns or, in other words, the acceleration of acceleration of technological or techno–economic change, including the potentialities of nano-, bio- or cogno-technologies, severally and collectively (Kurzweil 2005).

Certainly, it is at least under threat, if not altogether doomed as so often proclaimed, if there is even a slight modicum of truth in the claims of *post-, ultra-, meta- or trans-*humanists, a small but most vocal, prolific and publicized cohort now spreading throughout the world, well beyond the confines of Silicon Valley and California itself, and which now includes mainstream economists consulted by the Pentagon and analytical philosophers directing institutes at Oxford University and sympathizers who meet at Singularity Summits in Stanford. Indeed, a report published under the joint imprint of the National Science Foundation (a prestigious organization of American scientists) and the Department of Commerce of the United States a few years ago bore more than a few touches of transhumanist concerns, as was widely noted, especially by hostile commentators.[17] And a number of conferences have been held at leading US universities on these themes (and not sponsored by centres of the humanities and the social sciences, either).

To be sure, the idea of non-terrestrial, independent 'alternate humanities' has stayed in the West, more or less continuously, at least since the Copernican revolution or rather the Giordano Bruno reading thereof, which launched the career of the vision of the 'plurality of worlds' in modern times, possibly teeming with life and other kinds of rational beings. In fact, though this is often overlooked, in classical antiquity such speculations were not that uncommon, despite the conventional wisdom regarding the deep aversion to thinking about infinity amongst the Greeks, with a bearing even on the course of Greek mathematics.[18] It is explored today in such serious–amusing fields as 'space anthropology' (a field of inquiry which used to have its own journal), currently devoid of known objects, and even 'exo-biology' or 'xeno-biology'.[19] These fields also lack proven entities of study, despite all the efforts of the Search for Extraterrestrial Intelligence (SETI), a project cancelled by the US House of Representatives in 1993,[20] or the Search for Extraterrestrial Life (SETL) or Communication with Extraterrestrial Intelligence (CETI) or Messaging to Extraterrestrial Intelligence (METI), though the concerns that inspired such grandiose projects continue, especially in the efforts at detection of pertinent signals from deep space.[21] Various images, formulae and even a special logical language, LINCOS, or '*lingua cosmica*', have been put forward as the means for initial communication between earthlings and extraterrestrials and much more in the sphere of argumentation and speculation by scientists or scientifically trained journalists and other inquirers. Amateurs have also been enlisted in some numbers if only for the aggregation of computing power, on the 'grid' principle, tapping the idle computing power of thousands or tens of thousands of computers (even laptops, which of course have had, and are having, their computing power increased constantly) to secure the equivalent of the capabilities of supercomputers, a boon to academic scientists, and finance has been forthcoming from billionaires. By analogy, the study of as yet non-identified, but nomologically possible entities (i.e., entities whose existence would be compatible with the laws of nature, or the laws of physics), might even be said to some extent to be already in the pipeline along various causal trajectories, engaging the strenuous activities of the leading technoscientific 'inventors of the future', inventors of the beyond-the-human future, varieties of the post-human, which could be called 'ultra-anthropology'. Some of these appear likely to encounter at least prototypes or plausible prefigurations of their objects within the foreseeable future. Examples from within the realm of extraordinary entities, include the claim, 'I am a cyborg', in the title of a recent book by a scientist, the only self-proclaimed one so far, as far as I know, which may well be followed in due course by others of the same transgressive-prescriptive-propagandistic genre, such as, 'I am a chimera', 'I am a robot', 'I am an android', 'I am a bionic human', 'I am a cybrid', 'I am a __bot', 'I am a conscious computer' or, in an even stronger version, 'I am the Net, and I have become conscious', or 'I am a machine that has become alive'(Warwick 2002)!

To be sure, it is often not clear whether claims of this kind are being advanced with the utmost seriousness, but the basic point is serious enough. The advancement of the cause of rapid, directed species change, of radical human-directed or at least human-initiated, evolution of organisms and robots ('evolutionary robotics'), through whatever technological frontier looks most promising, through genetic/genomic engineering, synthetic biology (the possibilities of 'reverse engineering' in the lab, with commercially available inputs – of highly dangerous viruses, for example – have been demonstrated already), neuroscience, robotics, nanotech and so on. Humans are now prime targets for technological interventions by human beings, not just of remediation (as in deficiencies of all sorts, various kind of blindness or deafness, remedied with bionic sensors) or even amelioration, but of radical change, at least potentially and cumulatively, certainly in overall intent. Today, all these fields are increasingly cross-implicated, pervaded and severally accelerated by information and communication technology, even more than other cognitive pursuits, the convergence and synergy of current technologies, especially in the NBIC (Nano-, Bio-, Info-, Cogno-) complex of technologies, and their overlaps and interfaces or hybrids in a *combinatoire*[22] of great potential import ('bionano', neuroinformatics, bioinformatics, neuroelectronics, the production of neuroceuticals and nutriceuticals, nanorobots, the merger of genetic codes and computer codes, etc.) being one of the master themes of our time. The *disunity* of the sciences today as perhaps an irremediable condition has been a favourite topic of late amongst philosophers of science, against what was for a long time the standing presumption of the methodological and substantive unity of science, that is, the unification of scientific knowledge, perhaps in terms of a single fundamental master theory such as the Theory of Everything (TOE), at least as a regulative ideal, to be pursued indefinitely even if complete success could not be expected (Dupré 1993; Galison and Stump 1996). For some this was a project in the succession to that of the *mathesis universalis*,[23] which enjoyed its heyday at the dawn of the 'mechanization of the world-picture', of logic, reasoning, language, at least as a project, in the seventeenth century, a truly premature one, though today the main vehicle of the mechanization of the world, logic, reasoning, language, is not logic in the wake of the Fregean turn, but digitalization, indebted to the Boolean algebra of logic, to be sure (Dikjsterhuis 1961). Contrariwise, even if we may be resigned to the disunity of the sciences, the students of technology today rejoice in the current processes of the unification of *technologies*, one of the senses in which a pan-technological world is emerging, largely indebted to information and communications technology or technologies, to advances in computation, and to informationalist perspectives in every field of scientific–technical endeavour (digital physics, digital biology, digital archaeology, etc.), and beyond (digital philosophy, computational philosophy of science), though some critical philosophers of technology have long proclaimed and feared this very condition.[24] Some argue that the unification of the sciences today can only take place through the ongoing and unprecedented *convergence* of technologies, one of the most fateful megatrends of our epoch, specifically biotechnology and nano-, cogno- and info-technologies. Where once it had been argued, for instance, by the historian of science, Derek de Solla Price, that the *continuity* of science over time is secured to an important extent by technology – the instrumentation of observation, measurement and experimentation – not by the growth of theoretical knowledge (in part a response to the incommensurability debate in the post-Kuhnian philosophy of science), now it is argued that its *unification* will be brought about, not by axiomatic construction of theories of the widest scope and greatest ontological depth,[25] but by the unprecedented technological convergence taking place in our time (Bainbridge 2007; Martins 2005).

The tragic history of 'pseudo-speciations' and compulsory ideologically or politically driven imposed racialization in the twentieth century – the Holocaust, genocides, ethnocides and democides – reinforced the conviction that the unity of humankind, both factual and normative, natural and even transcultural, should be treated as a fundamental *credendum*.[26] The biological unity of (terrestrial) humankind, has been the ground of panmixia and in part of ecumenization, but in any case, the postulate of the unity of humankind in general became central to liberal civilization and to the pursuit of the social sciences since the eighteenth century. If this postulate were to be breached (for the first time, effectively, in the history of the species), sharp, irreversible psychological and cultural gulfs would open up in amid sections of the new humanity. Thereby, the radical conceptual or semantic incommensurability theses, strong incommensurability, incommensurability without comparability, radical meaning variance or, at best, weak incommensurability, conceptual or semantic, allowing for comparability, across cultures and paradigms, which flourished among constructivists in the sociology of scientific (and *a fortiori* all other kinds of)[27] knowledge, and amongst post-modernists, would be realized. The issue of the breakdown of this epistemological unity of humankind is not, curiously, much addressed in a world in which the profession of respect for universal human rights has become so pervasive, and the condemnation of racism and racial discrimination universal – even if the discussion of the natural underpinnings of the human 'race' category and achievement differentials never goes away for long, and it is quite open and unashamed amongst self-proclaimed 'neo-eugenicists', and of the emergence of distinct human species as a successor to ours, contemplated with equanimity by some liberal eugenicists, or the defenders of market micro-eugenics. Indeed, the issues raised by the projects and visions under discussion drastically impact on the question of the unity of humankind *tout court*[28] without qualification, not just the biological unity, because they would at least in the foreseeable future divide humankind substantially into the *improved* (genetically, euphenically, prosthetically, orthotetically, bionically, cyborgically, fyborgically,[29] as with the provision of new electronic senses, or super-senses, e.g., of vision microscopic and telescopic, and super-limbs, exo-limbs or an exoskeleton with superior motorial capabilities, neurologically) and the *unimproved*, or less improved, sections in accordance with the fault lines of both wealth and knowledge-capital, power-knowledge, and technological capabilities at the time of the shifts into 'designed' or 'redesigned' humans, 'designer babies' (the title of many articles and books) or 'perfect babies', 'techno-babies' or 'cyborg babies' (Stock 2002; McGee 1996; Davis-Floyd and Dumit 1998).

These projects and visions of more-than-human versions of the human (the roughly 100,000 years-old sapient version, the only version alive at the moment as yet), or simply the *humanum*,[30] overlap in so many ways that it is difficult to discriminate sharply amongst typical system-transformative trajectories and the overall end-states of such processes.[31] What appears to matter above all for those who espouse these visions or share the matching expectations, if not directly engaged in inventing these futures, is not so much one or another version of the post-human as much as success in at least one of the versions of the post-human, the sharp break with the 'actually existing' *Homo sapiens sapiens* in one frontier or another, in one 'liminality' or another of the *humanum*. It is noteworthy that some of these projects were opposed by computer scientists, particularly Jaron Lanier and Bill Joy – who are also critics of the 'cybernetic totalists' – and whose warnings received much publicity as well as much adverse comment from fellow computationalists and digerati after their surprising publication in *Wired* magazine).[32]

Supporters of these visions go by many self-assigned labels, and over the last five or six years a number of young left-wing adherents, beyond the core of the exponents in the computation and information sciences – such as Marvin Minsky, Hans Moravec, Ray Kurzweil – have tried to provide a left-wing, politically radical or environmentalist/Green slant to them. The Californian counter-culture of the late 1960s and early 1970s has left a lasting imprint among many of the digerati, and partly accounts for the appeal of 'open source' and 'creative commons' movements, hence the predicates of 'left' or 'radical' in a number of the labels. The movement's self-assigned names include transhumanists (the name of the most comprehensive association of its kind in being today), anarcho-transhumanists, cyberpunks (of a kind), Extropians, Singularitarians, bioutopians or biopunks, Viridians, socialist-feminist cyborgs, left Extropians, radical futurists, cyborg democracy, non-anthropocentric personhood theory exponents, transgenderists or postgenderists, technoprogressives, democratic transhumanists, libertarian transhumanists, socialist transhumanists, neo-Nazi transhumanists, transtopians, nanosocialists, vegan transhumanist socialists, cryonics devotees, digital or biological immortalists, radical life extensionists, radical body modifiers, 'morphological freedom' advocates, technoGaians, anti-primitivists, and so forth. Supporters of these movements also lavish choice epithets on those who reject their visions: bioconservatives, neocon bioconservatives, technoconservatives, speciesists, meat(ware) chauvinists, substrate chauvinists, primitivists, bioluddites, neo-Luddites, enemies of reason, enemies of science, technophobes, anthropocentrists, humanists (a pejorative term in these circles, as with Althusserians, Heidegerians, some environmentalists), or rather 'humanists' without a prefix, body fundamentalists, and so forth, though some computer scientists proudly claim to be 'cyber-Luddites' themselves.[33] Mostly it is the defenders of the radical transformation of the human who have taken the initiative, and stipulated the terminology to suit themselves: dysphemisms for their opponents, euphemisms or nicer-sounding terms for themselves. Of course, they might also call them 'tree-huggers', 'nature-freaks', 'nature-mystics' or 'biophiliacs', because of their reverence or care or sympathy for organic life as an undesigned creation, undesigned by humans at any rate, though the term 'biophilia' was coined by an eminent biologist, E. O. Wilson. He calls himself a 'naturalist', a low rank in the pecking order of contemporary biology, as he pointed out (though Darwin called himself one). He coined biophilia to designate a deep-rooted human attitude, a legacy of the Pleistocene 'environment of evolutionary adaptedness', which set much of our biopsychological make-up, as the sociobiologists and the evolutionary psychologists never tire of reminding us (Agar 2004). This attitude, Wilson says, needs reasserting in the Age of the Sixth Extinction, a global biotic extinction, largely anthropogenic.[34] Among the more learned, the 'bioconservatives' are accused of being children, or grandchildren of Heidegger, drinking from Heideggerian teaching at least at second or third hand from the immediate 'children of Heidegger', that is – Hannah Arendt, Hans Jonas, Herbert Marcuse, Gunther Anders, key critical philosophers of technology, and their commentators, perhaps without knowing it (Wolin 2001). Be that as it may, this does not dispose of the arguments, quite apart from the circumstance that in fact the sources are many and varied.

In current American usage, there is a tendency to single out 'technoprogressivism' and 'bioconservatism' as polar opposites in this field of controversy, though the choice of terms for two global stances can be all too easily reified (and the latter term was not chosen by the upholders of the positions so characterized). This already puts them on the defensive, confused with, or deemed to be neatly coextensive with, some other kind of cleavage in basic attitudes between (organized) religion and secularism, or between left and right, liberal and conservative, blue states and red states in political/world view or 'values' terms, derived from recent US presidential election campaigns. In any case there is a strong emphasis on bipolarization of stances and mutually exclusive binary categorizations.[35]

This agonistic, bipolarizing posture may be partly the result of the imprint left by the 'culture wars', 'science wars', 'Darwin wars', and so forth, of recent decades which, however, one might regard as a very inauspicious precedent if the ideals of the quest for truth, cognitive rationality and civility are prized (Brown 1999). It is not clear to me why the characterization of the rival views and attitudes as entering into intellectual conflicts as 'wars' is insisted upon, though in the past, analogous intellectual conflicts over matters of substantial import were termed 'debates' (e.g., the Socialist Calculation Debate, which raged over a couple of decades), or 'disputes' (e.g., the Dispute over Value-freedom, or the fact-value dispute in the philosophy of the social sciences, the Positivism dispute in Germany in the 1960s between Popper and Adorno), 'quarrels' (the Quarrel of the Ancients and the Moderns rephrased as the Quarrel of the Moderns and the Postmoderns, or the historians' quarrel in Germany in the 1980s), or 'controversies' (e.g., the New Value Controversy in Marxist economics).[36] But if they are to be called 'wars', they could be phrased as the war, or wars, 'over the *humanum*' (over the redefinition of the human being). In some ways, it might be argued that the war over the *humanum* is more consequential than the others, since it goes even deeper, though all these wars are very serious. Be that as it may, war over the *humanum* has not yet shaken academia in anything like the scale of the others. But then every human being has a stake in the outcome, even if they mostly abstain from participating in this war. Perhaps that is why academics are not so interested in such quarrels of less specifically academic import.

Our list of these visions, which are often jointly entertained by the same persons as if basically consonant, though they may be underlying tensions in some cases between the individualistic and the holistic strains, the 'libertarian' in the American sense, or anarcho-capitalists, and the green, 'deep green' (Sylvan and Bennett 1994) or 'deep ecology' currents, for example, is as follows:

(1) the neo-human
(2) the meta-human
(3) the trans-human

The Neo-human

This path would comprise essentially all the biological mechanisms, euphenics and eugenics (positive eugenics, though negative eugenics is not in fact excluded), for the substantial amelioration of the biological basis of human life, especially via genetic engineering whether somatic or germline (in addition to the ongoing Longevity Revolutions, the scientifically grounded immortalist projects, though, as Auguste Comte argued in one of his sociological theses that has come to acquire an unexpected relevance and topicality, a substantial alteration in the duration of human generations, from whatever causes, would transform the conditions of existence of human society) and more particularly the eventual transformation or bifurcation of the species into two successor species, of *trans-speciation*.[37] The eugenics in question is no longer state-driven, but market-driven, no longer principally concerned with mating or breeding rules and incentives or disincentives to procreation, like classical eugenics, but with the choices made by parents in the emerging 'genetic supermarket'.[38] The choices could also possibly be made by a 'Gamete Marketing Board', or global genetic hypermarket, not just of mates with the most suitable genetic endowments on Earth, if need be from sperm banks (whose supply has become increasingly international, if very uneven in terms of the nations of origin), or egg banks (which have been spreading rapidly in the last decade) (Baker 1999). The goal would be to meet the increasing demand for high-quality oocytes, and of new synthetic genes, or 'super-alleles', and the appropriate techniques for germinal genetic and genomic engineering, for the improvement and, indeed, practically the 'invention' of their progeny, according to their own criteria, generation after generation, though as generations are getting longer and longer. There may also be problems arising out of the coexistence of so unevenly genetically–euphenically–neurologically improved generations. However, the prospects of genetic self-improvement in one's own lifetime, of rectifying and improving one's genetic identity, also beckon, not to mention the prospects of indefinite tissue renewal via 'regenerative medicine'.[39] Some scientists envisage the outcome of these processes of market-driven gene-centric neo- eugenics, 'liberal eugenics' or the Republic of Genetic Choice (Agar 2004)[40] as extensions of the already-existing Republic of Choice, of consumer choice paramountly (and, of course, as in the Becker model of marriage and the family, children count as consumer goods).[41] It is noteworthy here that a recent, puzzling trend in American upper middle classes to have more than three children, has been tentatively explained by some demographers by the circumstance that kids are now viewed in such social circles as high-status consumer goods, as a brand of conspicuous consumption. 'Reproductive autonomy' is enshrined therein, drawing on expanding markets for gametes, offering the best genes money can buy, and perhaps 'morphological freedom' (the freedom to choose one's bodily appearance in every respect, zoomorphically for example), which could also be called 'morphological autonomy', in which each person or couple chooses the genetic profile of their progeny so that procreation is replaced by design or invention of future beings, as the break-up of the biological unity of humankind, at a time of ever greater panmixia, going even further than classical eugenics.[42] Not concurring with the socialist eugenic visions of the overall genetic improvement of humankind as a whole, this would result in the division of the human species into at least two successor species, a genetically improved or gene-rich species and a gene-poor species – as advocated by at least one geneticist, with surprisingly little adverse comment, in which the superior species becomes vastly more intelligent than the unimproved species (Silver 1997). Other writers offer the prospect of a succession of neo-human species on Earth, palingeneses without end for the foreseeable future, as the irreverent theoretical astronomer and cosmologist Fred Hoyle suggested in one of his essays on the long-term human future (1964).[43] It is a vision that comes close to that of the bioethicist Joseph Fletcher, in which he pointed out, several decades ago, the obvious but usually ignored circumstance that the much-vaunted enhancement technologies of all kinds could, in principle, be used in the opposite way, to diminish and downgrade human capacities, bred to serve, and thus produce 'para-humans', who could fill the place of servants or serfs or indeed humanoid pets (a prospect he did not find particularly distressing, though at the time he was a Christian theologian), at least until robots with sufficient motor and cognitive capabilities to match those of the para-humans (what one might call, in this context, and not necessarily in morphological terms, para-humanoid robots) came along to perform the same roles, and then they would be rendered superfluous (though some lingering sentimentality might keep some of them going as pets).[44]

The Meta-human

Under this heading we have in mind the visions that emphasize the emergent properties of the being, super-individual entity or psychocultural whole, which goes by the names of metaman, superorganism, noosphere, collective intelligence, world brain, hive-mind, hyper-networks, hyper-brain and so forth. The emergent whole that arises with the increasing connectivity and hyper-interactivity and hyper-connectivity, as it is often put, of humans through and with ever more ubiquitous and powerful computers/microchips in an expanding circle around and throughout the planet and through various thresholds of potentiation: 'We are all connected now', participants in the inclusive IT ecumene. Or: To be, is to be connected, to be electronically that is. The humans in question are, of course, increasingly 'enhanced' (genetically, neurologically, with electronic prostheses endowing us with super senses). This would also encompass things, because on the one hand all 'things', all inanimate things tend to become artefacts, everything 'raw' tends to be processed, all brute things subject to control by electronic communication, thus endowed (so to speak) with a modicum of animacy, networked, participating (in a manner of speaking) in the Web, or the 'datasphere': the so-called Internet of things, in which 'things', that is, intelligent things (sufficiently intelligent anyway), intelligent little machines of all kinds, 'smart buildings', 'intelligent houses' run with Bluetooth technology, and so forth, or at least things endowed with microchips, and through radio-frequency identification (RFID), too. All non-humans in which we have an interest, too, would be brought into a single planetary network of networks of constant information exchange and communication. In principle, not only the geosphere, but also living beings, the biosphere as whole, would also participate in the human-engineered, electronic planetary datasphere. The Earth would have its own common sensorium, as it were, albeit one controlled and monopolized by one species. It is difficult to resist the sense that we are witnessing a world that may be characterized as one of *technological animatism* or technological animism (as some commentators have already noted), in which, through ICT and the wonders of microsensors and microchips of varied kinds, all objects partake of some of the properties of self-movement or at least receptiveness and communicativeness that animists or animatists used to impute to them. It may be recalled that, for Durkheim, society comprised both individuals and 'things', that is, artefacts, machines and the entire anthropogenic world.[45] This definition, rarely taken seriously by sociologists, has never been more felicitous than today. Of course, 'things' in this sense are being sponsored by AI ideologues as candidates to mentation, individuation, a kind of agency, controlled evolutiveness, personhood, increasingly partaking of some of the properties of human individuals; sex, nonreproductive sex, presumably, between humans and robots has been solemnly predicted, not later than 2050, and indeed marriage, with all the privileges and practices that may come with it. UK Chief Scientific Officer Sir David King, has already warned that we must be prepared to concede an increasing range of rights to intelligent robots in the coming decades: moral, legal, political rights or, in other words, they will have legitimate, and perhaps irresistible, claims on us, and they will not be simply vehicles of our wants and objects of our projections or fantasies.[46]

The term 'metaman' is due to Gregory Stock (1993) ('metahuman' seems preferable, for obvious reasons), though this author has more recently advocated, above all, the path of genetic enrichment of humans, through somatic and germinal genetic engineering. Pierre Lévy's '*intelligence collective*' (1994), a kind of *logos* for the Digital Age (as our techno-bodies are bodies for the Digital Age), invokes the unicity of the 'active intellect', or 'agent intellect', a concept of the Aristotelian metaphysical tradition, particularly emphasized in the theological or philosophical anthropology/pneumatology of the classical Muslim Aristotelianism, which flourished in the twelfth and thirteenth centuries. Averroism has been, from a Christian point of view, a constant heresy in Western thought.[47] Durkheim, for example, has been recurrently accused of it, from the time of Scheler to the present. And indeed, for him concepts and *a fortiori* the categories of thought, are social, the common possession of social beings in a given society and in principle of all social or rational beings. Only sensations are individual and private.[48] Reason, the world of concepts and categories, is social through and through, and the growth of rationality involves their purification of local colour and contingencies; and science, he wrote on one occasion, is the social phenomenon par excellence. The body, for him, is only the principle of individuation (*principium individuationis*). Similarly, for Lévy and others of like persuasion, the world of cyberspace, increasingly the most commonly shared world, is the most inclusive common world in which all data, all information, all knowledge, increasingly digitalized, is deposited, stored, produced, processed, communicated, ideally in a global commons, and at any rate to a substantial degree. It is increasingly the medium of knowledge-work, rationality, science. Science, yes, the sciences *in silico*, or 'sciences of the artificial', like artificial intelligence and Artificial Life, obviously, but all sciences tend to become to an increasing extent sciences *in silico*, cybersciences, or at any rate there exist programmes in that direction even of Artificial Physics, with all the more justification if we take our physical world as itself adequately represented by computer simulations, if only out of sheer necessity, but to be a computer simulation in its own right, so to speak.

The idea of a World Brain is similar, even if not directly inspired by what is often called, loosely, Averroism, or the notion that mind, or at least the rational intellect, is a collective noun; Durkheim for example was charged with Averroism in this sense by Max Scheler, and has continued to be so charged by Catholic critics.[49] One of the first formulations of the idea of the World Brain to receive wide notice was due to H. G. Wells in a short book commissioned by the Institute of Intellectual Cooperation under the aegis of the League of Nations.[50] The World Brain was not a singular entity but rather the aggregate of, so to speak, all data banks (libraries, archives, museums, etc.), everywhere on Earth, arranged, and this was the crucial and the most problematic link in the argument, so that all parts of this universe would be fairly easily accessible to all the others, despite linguistic barriers and the distances involved; faxes had already been invented, though were not in widespread use until the last decades of the twentieth century, but microfilms were available. Today, the new ICTs enable this vision to a considerable extent, though there is a price to be paid for digitization and the fears regarding electronic preservation are quite legitimate and should be taken seriously by the fanatics of e-knowledge with their horror of paper. The attempts to capture all tacit, informal, unwritten knowledge of, say, experienced doctors, in software program form through the methodology of 'expert systems' has scored many successes, though surely the dream of capturing all tacit knowledge, all informal heuristics, through digital algorithms is just a dream, if only because knowledge in action generates or reproduces tacit knowledge outside these algorithms; tacit knowledge remains an irreducible category, even though any particular item of tacit knowledge or skill, can, in principle be explicated formally, so as to be inscribed as digital information, as a software program.

To speak of a World Brain, of course, does suggest a singular entity, and some recent speculators, even if with scientific training, have dared to articulate this vision, which could perhaps be better formulated as the idea of a World Mind, as there is no apparent proximate organic substratum for what they have in mind. At a lower level of analysis, it is organic human brains in the aggregate we are talking about, although brain–computer interfaces will play an important role, as well as bionic senses, unless we were to posit autonomous, rather than auxiliary, artificial intelligence. Still, it is not fashionable to speak of a World Mind, although to speak of a World Brain is more in accord with the naturalistic temper of the times. If we take the identity of brain and mind seriously, it should not matter, but even so some people have scruples about using the word 'mind', let alone 'spirit'. As to the concept of 'superorganism', the term will surely evoke the name of Herbert Spencer, and justifiably so, for what is meant by some scholars today is fairly similar to Spencer's conception, insofar as a collectivist-seeming term denotes in fact a form of life directed towards increasing freedom and diversity.[51] The superorganism's 'nervous system' is identified as its global capacity, the given society's capacity, for information processing (storage, control). At the limit we are talking about a world society. The global superorganism, in an 'evolutionary-cybernetic model of the emerging network society', is coming into being with 'the creation of an intelligent global computer network, capable of sensing, interpreting, learning, thinking, deciding and initiating actions: the "global brain"'.[52] This implies of course a growing global consensus about values and standards, a contingency that is rather a matter of faith than of reasoned prediction, especially as coercion is ruled out as the instrumentality for bringing about such a world society.

As to the concept of *noosphere*, this was formulated jointly in the 1920s by two friends, the mathematician and philosopher Édouard Le Roy and the theologian and palaeontologist Teilhard de Chardin, though it is associated today mostly with Teilhard's name, with some justification, as Teilhard went on to elaborate it in a number of works spanning several decades.[53] The inspiration for the term noosphere came from lectures given at the Sorbonne by the Russian geologist, Vladimir Vernadsky, where he introduced the concept of biosphere. Although very conscious of the increasing geogonic role of humans, and as a materialist and socialist, stressing the radical potential of human transformation of humanity's conditions of existence, he did not posit an extra realm of human ideation and cognition, capable of self-action and surpassing the material substratum of human life. Teilhard's noosphere, the realm of ideation and its material realizations, was constantly growing as part of an evolutionary cosmology. The human *totum* was in part its product[54] and, increasingly, the vehicle for is realization. The telos was the Omega Point, a key to the process of Christification of the cosmos, though the idea of the Omega Point, equated with that of the Singularity, or rather the Vingean, or technological, *Singularity*, has become a key concept and symbol of transhumanists who do not often share a traditional religious vision or affiliation. It probably was Tipler's work in computational anthropology and theology (1994) that drew the attention of transhumanists to this Teilhardian concept. In regard to Teilhard's legacy, he was one the greatest exponents of the cult of technological sublimity in the twentieth century; he regarded the invention of the atomic bomb more as an extraordinary, superlative manifestation of human powers, the supreme techno-fact, rather than a disaster for humankind, even though Oppenheimer, who led its construction, spoke of the physicists having known sin thereby. He situated the advance of science and technology, the key to cosmogenesis of redemption, in the context of a theodicy that resembled that of the agnostic Spencer, who prophesied 'the evanescence of evil' as human evolution, the supreme phase of cosmic evolution, proceeded. This is something that lives on amongst the transhumanists, though for them it is perhaps going beyond good and evil that really counts.

The Trans-human

Under this heading we have in mind the focal concerns of the trans-humanist movement (subsuming such older variants as the Extropians), which international movement, while not refusing, but endorsing, (1) the neo-human, and to some extent even (2) the meta-human, or cryogenics and other immortality projects, addresses particularly the issue of the *symbiosis*, a term launched in these contexts by the famous 1960 paper of the computer scientist J. R. Licklider.[55] Therein, Licklider used the phrase, 'man-machine symbiosis', the kind of machine he had in mind being electronic computers, so what he envisaged was that humans as users of computers, and the computers in human use, should pursue *mutual* adaptation rather than unilateral adaptation of one to the other, with the development of suitable interfaces, or a *fusion* of humans and machines. This was a version of the 'fusion of flesh and machines', the organic and the inorganic, which has been as much an artistic theme, among the Futurists particularly, as a technoscientific or science-fictional one. A fusion could perhaps take place via the upload of the contents of human mentation, or 'whole-brain emulation', in superior intelligent machines, endowed with autonomous intelligence, if not in their current biopsychological version, at any rate in a more bionic one.[56] Intelligent machines, or human brain-minds and intelligent, super-intelligent or hyper-intelligent machines, 'Avogadro machines', or inorganic minds, raise a frightening spectre. At the limit is 'Artilects', in the expression of the scientist Hugo de Garis, who constantly warns us to prepare ourselves for the emergence of such beings (to whose invention he has devoted his life), on the grounds that they will want to take over from us, and possibly even eliminate us, having undergone a mix of Lamarckian and Darwinian evolutionary processes. Notably, only the artificial world undergoes this mix of evolutionary processes. According to the Darwinian engineers, the natural world is exclusively Darwinian per the orthodoxy (de Garis 2005). Some of those who originally embraced Darwinismus only subsequently joined the transhumanists (such as the bioethicist Joseph Fletcher, once a Protestant theologian at Harvard, more recently an atheist), once more demonstrating the overlaps amongst these currents of thought.[57] The overall world picture this movement entertains is that of a cosmic evolutionary process in which the most intelligent species (on Earth at any rate) completes its vocation by its self-engineered, ever tighter association and even fusion with and eventual surpassing by such machines, entering a post-biological, rather than merely a post-zoological or a post-human stage, on a macro-evolutionary trajectory critically assisted by nanotechnological molecular manufacturing, towards the 'Rapture', the Vingean Singularity (often abbreviated simply as 'the Singularity'), or the Technological Singularity, and the Teilhardian Omega Point, often now simply the Omega Point (these are only some of the terms in the rich special transhumanist lexicon, though they are now entering the mainstream) (Alexander 2003).

Even leading scientists, in no way identified with the general thrust of the transhumanist movement, and generally regarded as quite respectable or 'sound', wary of going beyond the normal discourse of public science, argue that fusion with the (hyper-intelligent, or ultra-intelligent) machine is our only survival option on Earth. The physicist Stephen Hawking makes such an argument, as does the neurologist W. H. Calvin (2004), although Hugo de Garis foresees a war between even enhanced humans and the Artilect (superintelligent machine), even as he is engaged in its construction. Some believe that the ever-closer interrelations, mutualities, amalgamations of humans and machines will lead to a post-biological, discarnate human or post-human condition in some fashion. The Astronomer-Royal of the UK, and currently president of the Royal Society of London, Martin Rees, now Lord Rees, an establishment figure surely, *ex officio*,[58] so to speak, for example, argues thus (2003). It may be noted in passing that such distinguished scientists as Hawking, Calvin, Rees, de Garis and so forth. can hardly be dismissed as obscure cranks. Certainly, in such circumstances the classical conception of human technology (or technoscience) as essentially and necessarily prosthetic or instrumental, still advanced today by the more credulous enthusiasts, could no longer be held, if indeed such a conception has been warranted since 1945, if not earlier. In what sense indeed could the escalation of the lethality and the accumulation of the stocks of nuclear weapons be regarded as a continuation of *instrumental* technology? Only if the ends in question were extra-human – in the context of a theology of technology of sorts.

The general concern for the fast and substantial technologically enabled 'improvement of human performance' in work, sport, war (an ever-present concern), in scientific and other competitive achievements, and in general sensorial and motorial ways also.[59] This was shown very plainly in a joint report of the National Science Foundation of the United States and the Department of Commerce of the US federal government published in 2002. The report entailed that at some point human bodies will become increasingly 'techno-bodies' (not necessarily plugged-in, not necessarily in cyberspace), brains technologically implanted or enhanced, for example, through neuropharmaceuticals, starting with the endowment of a technological sensorium and technological motorium on the way to the crossing of some of the critical boundaries of the existing human bio-psychological condition.[60] It invokes a Human Cognome Project to follow the completed Human Genome and the ongoing Human Proteome Projects, and an 'engineering of the mind' or 'cognomic' engineering, to follow genetic, genomic or biological engineering in general (in fact 'biological' and 'genetic/genomic' tend to be equated, not only outside academia, but even within). This includes the provision of bionic sensors and actuators for, and cerebral implants in humans, to go with this new technological sensorium and motorium.[61] More specifically, it envisages a bionic sensorium and motorium, enabling the successors to the current version of *Homo sapiens sapiens* more direct bionic-sensorial access to the microscopic and megaloscopic (Geary 2002). This is, in a way, a new turn of the 'oculocentrism' diagnosed by the historian of ideas, Martin Jay, as prevalent in the modern West, or at any rate constantly denounced in French philosophical thought, more recently in Bachelard, Sartre and Foucault. It thus deracinates the human from the mesocosm, the middle world, it has dwelled in, as its natural habitat or *Umwelt*, as well the theatre of its *Wirkwelt*, even with all the scientific 'scopic' instrumentation developed and generalized since the seventeenth century, from the classical optical telescope or microscope to the latest devices that will enable us to see atoms. This involves, in principle, a kind of transcosmicization of our species, which might be seen as yet as another instalment of Piagetian decentration, the point of the exercise, or at any rate a substantial correlate of it, 'the effecting of all things possible', in the words of Francis Bacon, in the shape of transspeciation, the ingression of a new successor species, or more than one.

The question that has never been properly put to the general intelligence of humankind (or indeed to the citizenry of any nation), the moral Parliament of Humanity (with appropriate virtual representation of future and past generations, for at least half of the members of humanity are dead), is whether the substantial improvements of human performance in this fashion, turning every willing human who can afford it, for the market is supposed to decide in the end, into a kind of all-round super-athlete in body and brain-mind (with suitable optimizing brain implants, somatic genetic engineering, and more), should become the commanding goal or even a major priority for a planetary civilization, even a planetary business civilization.[62] The same might be said, of course, *mutatis mutandis*,[63] about all the techno-gnostic mega-projects referred to in the previous paragraphs, targeting one or more if not all the basic structures of human existence, deemed as mere negativities, ontologically speaking – particularly corporeality, vitality (organicity), mortality – which have accompanied the growth of human knowledge so far, but which the growth of knowledge, according to these movements, renders obsolete and indeed obstacles to the further growth of knowledge today and in the future. It often seems that, for them, knowledge, understood exclusively as *scientific* knowledge in its modern highly technicized sense, the pursuit of complete and total scientific-technical knowledge/reengineering of the world, is seen as the supreme, *overriding* endeavour, trumping, if indeed not removing completely, all other concerns for humans, or at least *quality* humans, and their legitimate successor species. The choice of a legitimate successor remains an important matter in this domain, as in much more mundane affairs.[64]

Not on the way to the Absolute Spirit, the identity of subject and object, with its Western Marxist version, as for example in G. Lukàcs, and others, or metaphysical historicism, the thesis of the radical historicity of the human being, which envisaged radical self-transformation of the human spirit in history, in a meontological perspective, or in the manner of the more tangible project of a self-determining humanity (as many sociologists envisioned, in the wake of Comte, as the central concern of sociology as the integrative social science, and in its time was even the 'LSE tradition', for example, but to what one might call the 'Absolute Epistemic Subject') (Martins 1993; Jay 1984; Fackenheim 1961; Ginsberg 1934). The Absolute Epistemic Subject: the epistemic would be strictly equated to the technoscientific, excluding therefore any noetic, contemplative, 'intellectual intuition'-type modes of cognition, at any rate an inorganic epistemic being seeking absolute all-embracing knowledge in (or as) an electronic–mechanical vehicle of very high, better-than-human, intelligence, whose habitat will not be the Earth, or any specific place, and which might indeed roam the cosmos in the quest for the most complete knowledge of the world, until it dissolves or dematerializes into pure light, as Bernal (1929) put it or, as we might say, into pure information, as formerly, in some cosmologies, beings dissolved into the Ultimate Reality of pure spirit. Some ethicists, as well as a number of scientists, have argued recently that a paramount duty, indeed the supreme duty, of human beings is the pursuit of scientific knowledge, the maximization of such knowledge, operative knowledge (not just one of their duties, even a high-ranking one at that) from which they also deduce that it is the strict obligation of parents to maximize the genetic capital of their progeny, providing them with all the genetic enrichment available in the current armamentarium of genetic/genomic/reprogenetic technology, so as to provide their descendants with the highest levels of general intelligence and of cognitive competences all round; all other good things will flow therefrom, it is assumed or implied, if not always explicitly asserted and elaborated. In more general terms, as we have noted, it has been argued by the evolutionary biologist John H. Campbell (1995) that it is our strict *duty* to assume the conscious direction of human evolution in the first instance and, insofar as it is within our power, of the evolution of life. It is not just a question of possible worlds, of desirable worlds, thinking in the optative, of alternate scenarios of the human future to be pondered, but of an overriding moral *imperative* to control evolution and to steer it towards the maximization of knowledge and power, or power-knowledge. Already in the 1960s, when the potentialities of genetic control were becoming clear, the sociologist and demographer Kingsley Davis exclaimed: 'We shall become like gods!', enjoining us not to miss this cosmic opportunity.

The ongoing changes in the computational thought-world involve the conceptual recognition, a categorical shift, in some ways a restructuring or overcoming of folk taxonomies, naïve ontologies, or 'our [common sense] conceptual scheme', of vitality, animacy (Artificial Life), mentality (Artificial Intelligence) and personhood (Strawson 1959). Artificial Persons, nonhuman persons generally, want to extend the well-established entitlements of personhood.[65] Due to Christianity and the Enlightenment, this extension is to various categories of nonhumans, even as they exclude some categories of humans from the universe of persons, so that the universe of human beings becomes only a subset of the universe of persons, and the universe of persons extends beyond that of human beings.[66] They even implicate the issue of the putative or potential soulfulness of their 'bots': the 'theology of robots', an investigation into the question whether artificial beings may become ensouled (thus Artificial Souls, as we might call them, by parity of denomination with Artificial Minds, Artificial Intelligence, Artificial Persons, Artificial Societies), so far only pursued, as far as I know, by Christian theologians, though there should be no reason why, for example, Shinto believers or practitioners should not ascribe spirit-status to them, not to mention the long-enduring Golem tradition in Judaism, once evoked by Norbert Wiener,[67] has comprised one of the research fields pursued at places like MIT (the 'God and computers' project directed by the Lutheran theologian Anne Furst).[68] Thus they would be placed on a par at least with that 'moral considerability' typical of the members of the human species, and the granting of rights, moral, civil and even political rights to their creatures, or the most evolved variants thereof, should follow. And not just as fellow 'actants' (the Greimasian semiotic term, once confined to narratology, recently popularized by Bruno Latour to stipulate that we encompass agents and non-agents in human societies in the same tightly coupled domain, though the implication that we and our artefacts are all discursive items in an unfolding narrative without a narrator does not appear to have been followed through), but as fellow militants in the 'war against terror' (the number of American book publications on 'citizen cyborgs' as well as 'cyborg warriors', conscripts, not volunteers, conscripted through programming, one might say, is increasing).[69] This at the very time when the curtailment of the rights of human citizens (this locution was once a peculiar redundancy), even in the United States, in the name of 'homeland security', is all too likely, for alleged reasons of national, regional or global security against many varied, insidious modes of terrorism, through biological warfare or cyberthreats, the deployment of radioactive substances, as in the recent cases of polonium-210 poisoning in London and elsewhere (undetected at first by the scientific instruments deployed, such as the standard Geiger counters), not to mention the miniaturization and easy availability of sophisticated science-intensive weapon systems of every description and ever-increasing precision, lethality, and especially the capacity for inducing mass terror for every terrain and circumstances (particularly the nuclear question, the threat of nuclear proliferation continues unsolved) (Panofsky 2007). This has expanded with recent and ongoing advances in today's irretrievably decentralized, worldwide market economy (both legal and illegal) of sophisticated weaponry for a very wide range of niches (Rolo 2006).

In these visions the strain of 'technological gnosticism', a term coined by the American philosopher of technology Victor Ferkiss (1980), reaches perhaps its culmination, to the point of concluding that we are living in an emergent techno-gnostic civilization. At the very least, it could be argued that these demonstrations of technological gnosticism (which have become very self-conscious in recent years when variants of gnosticism have been embraced consciously and openly) by militants of biotechnological intervention looking towards the comprehensively 'genetically modified planet'[70] and other forms of technological intervention in biospecies modification or creation (a start has been made with synthetic bacteria), and the surpassing of natural, carbon-based intelligence (and indeed carbon-based life at large) by silicon intelligence or silicon life (though admittedly even silicon might be replaced by another physical substance, substrate or process). The ultimate object of these pursuits is a highly desiderated 'technological transcendence', transcendence of an atheological and supposedly unmetaphysical kind, in and through technology, or rather through a particular constellation of contemporary technological frontiers, above all through the technologies of 'strong' AI. This would be a wholly unfaithful mirror to our microchip-saturated, computer-ubiquitous, through the WWW of today to Web 2.0+ (or even to the already outlined 3.0, or the coming Semantic Web, and so on) 4G+ mobile phones, and an increasingly RFID-pervaded technological civilization. Adding to the Internet or WWW, semantic or otherwise, of humans, 'the Internet of things', where, as noted previously, every terrestrial object, and indeed every manmade object in outer space, already somewhat polluted by us with the thousands of satellites that have been placed in geostationary orbits since the 1950s (Bertell 2000), would be tagged, or at least could be tagged in principle, and thus in communication with every other object and potentially every subject or every putative information-processor, or actant, in the Latourian twist, too, within the planetary datasphere. To be an actant, or rather a microchipped communicant (for they do not necessarily act, except in so far as transmitting information, for which they are programmed, counts as action or actancy), is in this sense to be an integral part of the planetary datasphere, or of a kind of pansophic medium, if one were to equate, wrongly, in my view, the sophic with the informational in the sense of 'information' that occurs in the expression ICT. In current management literature, of all places, it is common to draw a fourfold distinction, in a definite hierarchy comprising data, information, knowledge and wisdom; one might, following this mono-hierarchical set of categories, discriminate among four spheres, with the datasphere, strict sense of data in this context, as the first and lowest, the infosphere (a term already in use, like datasphere) the next highest, the cognosphere (as we might call it), the third, the wisdom sphere the highest. Most writers on these matters are not that ambitious, and the very idea of a planetary pansophic sphere would seem metaphysical nonsense (as if informationalism itself were not without metaphysical commitments, sensical or not).[71] It has certainly precipitated them, as one might expect from the current view shared by many physicists, though it goes back to Norbert Wiener – that information has to be ranked with energy and matter as the three fundamentals of the physical universe, it is not just a basic category applicable only to the biological or the sociocultural realms.

There does seem to be a *prima facie*[72] case that such visions should be taken seriously and not, as the more glib or defensive or disingenuous might like to claim, as the ravings of a cybernetic lunatic fringe, or of a tiny minority of utterly unrepresentative, minor or marginal scientists, engineers, science journalists and commentators indulging in media-hype, of digerati and the 'artificial intelligentsia', 'cyber-totalists', who merit no serious attention, and whom we need not fear (and may even laugh at). Or, as some like to say: 'It's never going to happen'. *Something* will happen on the way (unless one or another of various ecological calamities or societal collapses, or even the fatal impact of astronomical bodies from outer space, so much stressed by a number of scientists in recent years, enjoining us to permanent vigilance, occurs first), so there is a good case for reviewing these visions and exploring scenarios associated with them, even if those not acquainted with the literature and the scientific voices therein may all too easily dismiss them as science fantasy or science fiction.[73]

Alas, the matter cannot be brushed aside so easily, indeed so flippantly and arrogantly.[74] I claim that it would be utterly irresponsible with regard to both science and society to try to ignore the potentialities and scenarios we have reviewed, far-fetched and fantastic as they may appear to those endowed only with tunnel vision or ineradicable Panglossianism.

We say 'science and society', in the conventional locution, though science *is* in and of society, as well as, in its truth-content, *out* of it, if only in another region of the Popperian world, that is, the realm of the products of the human mind, whose implications over time are never foreseen in full, and often unwanted and undesigned, while society, bound more and more to technoscientific conditionalities, seems to belong, as it were, to science, or at any rate to fall increasingly *within* the technoscientific way of world-making with all it portends, for good or ill, now or in the future, even if with an animalish or uncontrolled residuum, ontologically speaking, still to be domesticated, or perhaps one should say 'enhanced' or redeemed, *la société sauvage*,[75] as it were. It seems to me, that to speak of an interface, or interfaces, between science and society, as is common practice, involves an unsatisfactory mode of conceptualization and but a partial truth; science is *in* society as well as *out* of it (that has always been true), while more and more of society (and of the human being on the way to cyborgification) increasingly falls *within*, or is incorporated or embedded in, the technoscientifically made and remade world, virtual, virtual-real and real. This is the Great Transformation in science–society interrelations, accomplished through the interpenetration of science and technology, the technological intensification of science, the knowledge-intensity of key industries. Hence the knowledge-economy, the capitalization of knowledge, the emergence of technocapitalism, specifically one in which ICTs provide its defining or constitutive meta-technology. The term *interface* in the current literature of the cyber-theorists is often used, rightly or wrongly, in conjunction or even interchangeably with, the modish terms 'merger' or 'fusion' or 'symbiosis'.

As a number of scholars have pointed out, circumscribed local laboratory science has spilled over and expanded to encompass the world, the *world-as-laboratory*. Similarly, a number of *scientists* have argued in the last two decades that we are engaged and implicated involuntarily in a 'titanic planetary experiment' (they have used this phrase, too) through the cumulative and pervasive biospheric, atmospheric, geogonic impact of our material practices. In other words, if they are right, and as scientists they perhaps know what they are talking about, we may view our world as a *world-as-experiment*, an all-world experiment the outcome of which is highly uncertain, a planetwide experiment that may well go wrong, like common, partial infra-world experiments and, which, in any case involves an experimental situation of a wholly unprecedented scale. So much so, that a Nobel Prize chemist, Paul Crutzen, has called our epoch the *Anthropocene*, precisely on the grounds that within the *Holocene*, the epoch coextensive with the existence of humans, the period since the invention of the Watt steam engine inaugurated a phase of vastly increased and accelerating anthropogenic impact on our 'shrinking planet' or world-lab site. The term *Holocene* was invented before the transhumanist movement took hold, though at the time the term *Anthropocene* was coined, a few years ago, it was already flourishing, an instance of the late flight of the Owl of Minerva perhaps.

In any case, we live already in the science fiction of yesterday, as the biophysicist J. R. Platt, once expressed it (that statement was put forward three decades ago). To which one might add that, with the acceleration of acceleration in ICTs and other converging technologies, perhaps we will be living soon, not just in the science fiction of yesterday but also ever closer to the science fiction of today. Why read contemporary science fiction, indeed? We are reading it directly, at first hand, in the world we inhabit, though many of us are *digital immigrants* (as we have been called), born before the digitalization of the world, still something of the exotic for us, though an increasing number count as *digital natives*, born and bred in the digital world – not just the human realm, for the *Book of Life* and indeed the *Great Book of Nature* itself, now written in binary digits, as indeed all books *tout court*, with a small 'b', are becoming digitalized – totally immersed in it, and perhaps perfectly at home in it. And presumably every succeeding human generation from now on will comprise mostly digital natives, as the digital becomes increasingly the monoculture of the mind, and the world, perceived or thought, digital through and through.

Perhaps some will feel our stance must always be *pas d'ennemis dans la technoscience*, no matter what, as in: 'my country right or wrong', 'the Sovereign can do no wrong', or *pas d'ennemis à gauche* (none of these watchwords were invariably conducive to the greatest good of the greatest number, to put it mildly). *Fiat scientiae, pereat mundus*,[76] though of course only under some specialized though not wholly idiosyncratic interpretations of the potentialities of science and technology, especially the alleged supreme *duty* to pursue its implications of alter or trans-humanization to the utmost, even to the point where the concept of duty could no longer have any meaning. Certainly, the *mundus*[77] of natural humans, organic humans, fellow members of our species, are still made up, too many of them, largely of 'meatware' or 'fleshware', dwellers in 'meatspace', earthbeings. At some point, the surviving natural humans would no longer be able, in principle, to say of such beings '*mon semblable, mon frère, ma soeur*',[78] as we can now say, in a perfectly good sense, of and to every one of our coeval human beings on Earth, of all our species predecessors, and of our natural, undesigned posterity, as we have been saying since the Stoics and the rise of Christianity. It would shatter the unity of humankind (both a Christian and an Enlightenment teaching), the unity of human and nonhuman life (a key Darwinian teaching), the human bond with the biosphere, which has defined the Blue Planet, quite possibly a singleton, a *unicum*[79] (according to the 'Rare Earth' hypothesis) (Ward and Brownlee 2000). Perhaps this is the last *tremendum*[80] we could contemplate as humans. Though for some, even now, it would count, not as a calamity, not as a tragedy, not even as 'The End', not entirely to be regretted, of the most predatory and destructive species on Earth (as some radical ecologists have felt), but as 'The Rapture', which they long for now (Alexander 2003). They long for it, and work for it, no matter at what cost to most natural humans (bundles of meatware, no matter their bionic prostheses, living in 'meatspace', the lived, hodological space of our *Umwelt*,[81] even if in a life-world more and more technologically mediated), for salvation is not promised to all; only a few could or would be 'saved' to enter, irreversibly, the disembodied, inorganic, post-biological 'computational plane of existence' – or, eternal life, in this version – and mostly prescind from the biosphere (Smil 2006).[82] Still, many vanguards, sometimes wearing the specious mantle of reason and science, of knowledge and power, engaged in a messianic, perfectionist, redemptive mission, have come and gone in the last two centuries or so, their salvific promises exhausted. Not, however, without causing immense damage.

Notes

1 *Ex machina libertas* is a Latin phrase that means 'freedom from the machine'.

2 French scholars sometimes write of 'the sciences and technologies of information and communication' (STICs), in some ways a more satisfactory expression.

3 Indeed, Kurzweil speaks of 'spiritual machines' concerning software programs, as they replace brainwork, or permit much more versatile work on brute matter, and others

speak of 'dematerialization' with enthusiasm in this connexion, as if microelectronics, computationalization and digitalization would enable human beings to transcend the constraints of the material world. The current apotheosis of dematerialization via software, exalting the newest technologies of information and communication as the paramount vehicle of spiritualization, can be seen as a variation on a theme. Since at least the late eighteenth century, electric and electromagnetic machines have been perceived as more or less 'spiritual', or at least less grossly physical than mechanical machines or steam engines, for example. Even the phenomena of magnetism seemed, in the eighteenth century, to portend a break with the mechanical world view, and 'mesmerism', the teachings of Mesmer, which found favour among some Jacobins, such as Marat, became a social movement for social radicals critical of both the scientific–mechanistic and the religious pictures of the world, though in many cases it would be more accurate to say they claimed to be pursuing real science against the academic mandarins of the scientific establishment.

4 In current scientific usage *cyborgs* need not be even part-human, or of human descent; a scientist recently claimed to have devised a moth-cyborg. In fact, this was part of a project of DARPA to develop 'reliable tissue-machine interfaces' in the form of Hybrid-Insect-MEMS (these results of the intertwining of nervous tissue with silicon chips may be useful to keep remotely controlled spies in the battlefield and elsewhere, a bit like the proverbial fly on the wall that could listen in to the conversations of the great). Presumably, there is far more to come from this project in terms of what we might call *zoo-cyborgs*, the fusion of nonhuman organisms, insects or other animals, with micro-electronic devices to constitute new entities endowed with some artificial intelligence and a residue of organic life (especially neural tissue), though until now insects have been the privileged targets of these developments. So far, they fall mostly under the rubric of weapons of war, classical or against terrorists, but peaceful agroeconomic applications have already been thought out, though it is not clear whether the weaponization of insects will be abandoned. The term *cyborg*, as is well known, was originally devised as a contraction of 'cybernetic organism', and in its pristine version referred exclusively to phenotypically modified (and exosomatically equipped) rather than genetically engineered (or reengineered) humans, though the term *organism* provided for its extension to nonhumans. In practice, the first so-called cyborgs were conceived as *anthropo-cyborgs*, aiming at the fusion or integration of microelectronic devices and other equipment with the brain or the body at large, to enable space colonization by human beings not biologically suited to the conditions they would face in space vehicles and outer space. For some cultural theorists, subsequently, above all with Donna Haraway and her numerous epigoni and echoes, the cyborg underwent a kind of exorbitation and re-functionalization, left the world of NASA (it was two NASA scientists who coined the term), and became the vehicle, not for assisting humans in outer space, but rather for assisting in the project of the abolition of the human, or at least of the category of the human, the merely human, and certainly the gendered human, on this Earth. At least one scholar has proposed, instead of 'cyborg' (whose history, as we have seen, is quite tortuous), or indeed 'bionic human' or 'human with implanted technology' the softer term 'cyborganism'. By this he means someone like a writer at his or her computer console, 'enhanced or extended by computer technology, which is external to the body' (Lebkowsky 1997). If cultural studies scholars have been enthusiastic about 'cyborgs', analytical philosophers are still debating the question of 'zombies', intelligent beings lacking consciousness, and whether indeed we are conceivably zombies, 'consciousness' being a fiction like other 'folk-psychological' terms, according to a well-known school of thought. 'We are all cyborgs', 'we are all zombies': two slogans that have not yet joined up, being heard in different quarters of academia.

5 *Eo ipso* is a Latin phrase that means 'by the thing itself' or 'by that very act'. *Desideratum* is a Latin term for 'something that is desired'.

6 The boundaries between individual and individual, between body and body, between mind and mind, between brain and brain, are much less often adverted in this literature.

However, this theme is involved in the pursuit of the meta-human, discussed below, where the connectivity of all minds/brains, like the connectivity of all machines, and indeed of all beings with the proliferation of microchips attachable to every natural being, may turn the planet into a vast infosphere or at least datasphere, and the Earth will become its own sensorium.

7 'Morphological freedom', or transgendering, are still regarded or defended as rights, moral rights that should become statutory rights, not yet as *duties*, though some radicals who view androgyny as the optimal human condition, in the manner of Plato's myth, and some mystics, a condition that should prevail in the human world as far as possible, would certainly argue or at least imply that it is a matter of duty (not just concerning our conduct in gender-blind terms, but as transforming our own *being* and thus the human type as such: from gendered humans until now to transgendered humanity for all time).

8 A famous phrase of British Idealist political philosophy (esp. in B. Bosanquet) regarding the mission of the reforming democratic state.

9 *Ab initio* is a Latin phrase that means 'from the beginning'.

10 *In silico* is a Latin phrase that means 'in silicon', and is used to refer to something that is done on a computer.

11 *Outré* is a French term that means 'unusual' or 'strange'.

12 For examples of this genre, see writings by Dennis Gabor and Danny Hills.

13 *Esse* is a Latin term that means 'to be' and refers to being or existence. *Fieri* is a Latin term that means to 'become' or 'be made'.

14 *De jure* is a Latin phrase that means 'according to law'. It can be contrasted with the Latin phrase *de facto*, which refers to something that exists 'in fact', even if it is not prescribed by the law.

15 Even on the Gould–Eldredge doctrine of 'punctuated equilibrium' in macro-evolution, species creation or species selection after planetary catastrophes – the great accelerators of biological evolution, in their view – it would still take millions of years, at least on some understandings of the concept of biospecies. This is a somewhat controversial matter, as the bitter and protracted methodological disputes over biological systematics well into the second half of the twentieth century testify.

16 At the same time, one often finds claims that natural biological evolution has practically stopped (these claims were often found, too, in the nineteenth century), or is so slow (a somewhat different point, but the two claims are often conflated), and that humans have to assume the grand task of deliberately speeding it up and shaping it in conformity with some predefined aims. Four observations are in order in connexion with the premise: (a) natural biological evolution is most unlikely to stop, on well-established Darwinian grounds (it is unnecessary to unpack this argument here), though in many spheres it is 'too slow' (not by the evolutionary clock, only in relation to our aims and expectations, as fast has become the imperative in our lives); (b) the 'evolution explosion', or very fast micro-level evolution referred to, is and will go on taking place, accompanying the biomedicalization of the world, and the widespread use of antibiotics (here the complaint is that evolution is too fast); (c) the extinction of species, the disappearance of habitats and ecosystems proceeding apace, directly or indirectly, through multifarious human actions, which may amount to a Great Extinction on a scale only matched five times previously in the entire history of life on Earth, although this one is the very first to be induced to any substantial degree from within the Earth, and the first to be induced by living things, let alone by the actions and general conduct of a *single* species, while the others were due to the impact of astronomical bodies and take place at a greater speed than the previous ones, at least overall; (d) the transfer of exotic species through deliberate human agency or inadvertently, or the cumulative outcome of such deeds, and their side effects, taking place on a greater scale than ever before in the last few decades, with their destructive impact on native, and especially endemic, species. Of course, in Artificial Life, we can bring about artificial evolution, that is,

the evolution of digital organisms, through a mix of Darwinian and Lamarckian methodologies (letting variation take place and also guiding it).

17 One of the two editors of the report, W. S. Bainbridge, is clearly identified with the transhumanist movement.

18 This is a conventional view, though Spengler, who had a mathematical education, perhaps made the most of it in his historiosophic scheme in which Occidental or Faustian culture is marked by its obsession with and restless quest for infinity in every domain. One might well subsume under this rubric the Hegelian quest for self-infinitization, perhaps the highest state of metaphysical intoxication ever reached in the West, at least since classical times, which metaphysical intoxication, in a way, the young Marx shared, or so it has been ably argued by Robert W. Tucker (1961). True, numerous texts collated by the historian of philosophy, Rodolfo Mondolfo (1952), provide evidence that speculations about other worlds, infinitely many worlds in space and time, were not that uncommon in the classical world, mostly in connexion with atomism, for which reality consisted at its most fundamental level, of atoms and the void, empty space, and thus infinity, perhaps infinitely many worlds though one should not overlook the other major tradition of atomism in modern times, the Indian one. This infinitism, however, was never received in Greek mathematics and mainstream Greek philosophy. So, broadly speaking, the Spenglerian characterization holds, though Spengler, like many German scholars of his time traced Western infinitism to Nicholas of Cusa rather than to Giordano Bruno (cf., the great importance ascribed by Cassirer to the Cusan in his history of philosophy, where he is discussed at considerable length, much greater than in any other history of philosophy that I know of).

19 There are no perfect synonyms and preference, for one or the other term can be well-grounded.

20 Due to budgetary concerns, the US Congress cancelled the High Resolution Microwave Survey (HRMS) in 1993. The HRMS was a NASA SETI program designed to detect radio signals from extraterrestrial life.

21 At the same time, it must be said, as it cancelled the Superconducting Supercollider, the biggest project ever conceived by Big Science in the history of the world, and which will probably remain unmatched, at least for the foreseeable future, on the construction of which in a Texas site $2 billion had already been spent. It has been argued that this project (and perhaps the SETI one also) was cancelled largely because the end of the Cold War removed the necessity for America to constantly prove itself to the world as Number One in highly capital-intensive Big Science or even Megascience, as such projects are now called (especially the ones since the 1970s), and Big Tech, or what one might call Conspicuous Science, in the spectacular way of the Apollo Projects, culminating in the perhaps once-and-for-all landing of humans on the Moon in the Apollo 11 voyage in 1969. At any rate, it has not been repeated, and there are no immediate prospects for doing so, and indeed the whole Space Race between the USSR and the United States might be so characterized, a happy diversion from real war, and the gigantism of this venture in the pursuit of fundamental physics research. Michael Polanyi, an acute commentator on the Soviet economy already in the late 1930s, characterized it in the late 1960s as one in which 'conspicuous production' played an important role. We can now see what followed was a turn into deliberately changing the fundamentals of terrestrial life via agribusiness, genetic/genomic engineering, the molecularization, technologization, informatization, digitalization and marketization of genetics and eventually the brain/mind (the orthodoxy posits their identity, whether token identity in its weak form or type identity in its strong form) into superior versions, followed this shift away from cosmic knowledge pursuits. The Bayh-Dole Act came just in time to play a midwife role in this transition from the First Space Age to the Century of Biotech, from Cosmocracy to Biocracy. The Second Space Age, advocated by a number of transhumanists, and always kept in mind by them as providing an escape route for the emerging superior beings they exalt, has not really started.

22 *Combinatoire* is a French term that means 'combination'.

23 *Mathesis universalis*, derived from the Greek word for 'science' or 'learning' and the Latin word for 'universal', refers to a 'universal science' that is closely based on mathematics.

24 Both Jacques Ellul and Herbert Marcuse are examples of such philosophers, though in *their* case they were, and indeed still are, severely criticized for their alleged overstatements concerning the pervasiveness of technoscientific totalism or 'monism', as Ellul put it (1954).

25 The notion of ontological depth of scientific theories was particularly well explored by Watkins (1984), in the wake of Popper.

26 *Credendum* is a Latin term that means 'a thing to be believed'.

27 *A fortiori* is derived from the Latin phrase *a fortiori argumento*, meaning 'from the stronger argument', and is used to refer to a side or position that has stronger evidence.

28 *Tout court* is a French phrase that means 'very short', 'simply' or 'without anything else'.

29 A term was suggested by Gregory Stock.

30 *Humanum* is a Latin word that refers to the human.

31 The epoch of 'anatomically modern humans', the figure in the text being perhaps the most often presented: recently, significantly higher estimates for the duration of this epoch have been claimed, though the matter has not been settled according to a new scientific consensus amongst palaeoanthropologists, and for convenience we have adopted the figure in the text without commitment to its final truth.

32 J. Lanier, cybernetic inventor, computer scientist and musician, coined this useful expression. Note that he had previously put into circulation the expression 'virtual reality' in the context of electronic digital simulations, a locution he has adverted, he borrowed from the writings of the aesthetician and philosopher of mind Suzanne Langer (1953), wherein the terms 'virtual reality', 'virtual space' and 'virtual time' occur in the analysis of the epistemological status of our aesthetic experience of artworks in general, but especially perhaps of paintings. As is well known, 'virtual velocities' was an expression in mechanics at least from the eighteenth century (it got a mention in *Das Kapital*) and virtual rays in optics, while the Latin term 'virtualis' also appeared in scholastic philosophy.

33 It is not clear how many people subscribe to the anarcho-transhuman thesis. In their website, it is claimed that it militates against the tyranny of government (any government, as befits anarchism), against the tyranny of capitalism, and against the tyranny of genes. It is transhumanist in the sense that it affirms 'the desirability of fundamentally altering the human condition through applied science and technology' and, yet, unlike the Extropians, 'it does not believe in optimistic futurism' (Anarchism and Transhumanism 2015). Still, overcoming death is also part of their brief.

As for the chauvinists, all this is not spelt out, but I suppose they also mean to disparage those addicted to what is now called 'treeware' that is, paper (or indeed papyrus), printed matter, books, libraries, newspapers and magazines in print form, writing (except electrographic writing) or print media in general, insofar as it depends on organic matter: the end of 'print-capitalism', in the famous phrase of Benedict Anderson, characteristic of the age of European nation-states, as well as indeed of other early modern types of polity, the emergence of digital or e-capitalism and global civilization of a sort, perhaps, in which digital media capture or overwhelm analogue ones. Though plastic trees would be more acceptable, no doubt, not partaking of organic life, but at present they do not as yet lend themselves to the production of paper, or indeed to anything much of value (their disvalues are more apparent).

34 Although some deep ecologists are pantheists and many might lean towards panpsychism, these terms have not entered the vocabulary of these culture wars, though many transhumansits are radically hostile to present-day religions.

35 With a quite different connotation from the European colour symbols, where 'red' stood for the left and 'blue', at least sometimes, for conservatism or traditionalism. This American usage may be due to the circumstance that the common pejorative term 'redneck' (easy to abbreviate to

'red') stands for reactionary, country bumpkin attitudes. One of the key divisive issues on these lines was over abortion. But there are signs that the situation is becoming more complicated, and attitudes more sophisticated, and the simple division between red and blue states is no longer as sharp.

36 Many of these expressions first occurred as translations of the German word '*streit*' as in *Methodenstreit, Historikerstreit, Positismusstreit.*

37 The (First) Longevity Revolution, starting slowly in the West in the eighteenth century and accelerating in recent decades, initially was essentially due to better nutrition (most clearly shown in the rising average height of Western populations generation after generation) and urban public controls of sanitation, the completion of the 'demographic transition' or the shift to the 'modern', post-Post-Malthusian population regime with historically unprecedented low birth rates and historically unmatched low death rates. This is of course an ideal type, though Southern European countries in the last two decades have come to practically embody it, with medical interventions playing an increasing part in the twentieth century and even more so in the twenty-first so far. One might characterize the process as a whole as one of the increase of 'physiological capital' (as the Nobel Prize winner Robert Fogel puts it), good health, increased ability to work over a longer chunk of the total lifetime, rising life-expectancies with still a substantial period in which the quality of life may be high ('seventy is the new fifty'), which a number of economists and economic historians have argued should be taken into consideration in national income accounts. Or, in Western countries, in other words, we are richer than the official GDP per capita statistics tell us, by a few percentage points in the UK according to some estimates, though decreases in such physiological capital have taken place in the last few decades in the USSR and its successor states as well in Eastern Europe, particularly with decreasing life expectancy for males over the last quarter century or so in Russia (already apparent in the last two decades of the USSR). There are also other negative health statistics, and thus the course of 'technophysio evolution' (Fogel again), or the accumulation of 'physiological capital', need not be smooth and monotonic, and indeed can even be reversed. The coming of at least the currently hypothesized Second Longevity Revolution, with the eventual flourishing of innumerable supercentenarians, would be due to modern biotechnology, genetic engineering and 'regenerative medicine', as well as to more stringent environmental controls. This fact is, curiously, somewhat neglected among contemporary super-longevity or 'radical life extension' theorists. Recently, it has been argued that age should not be measured by years-since-birth, as in common practice and in law, but by mortality risk and remaining life expectancy. I am not sure whether official documents like identity cards and passports would record these estimations, though doubtless national statistical offices would be very interested for demographic forecasting as policy aids. More likely, as some radical feminist philosophers have argued, such documents should not record anything mutable (as a matter of technical possibility and of right), like gender, nationality, name, ethnicity, or even age (given rejuvenation techniques), which would leave very little to record. The British post office has suggested that everyone should have a unique number, as people move residences so much, what one might call the 'Zamyatin solution' (from Zamyatin's dystopia, 'We') and perhaps this could also be applied to identity documents: the philosophical questions arising from these procedures have not been properly addressed as such.

A recent inquiry by a biologist surveyed feasible paths to immortality, compatible with the findings of evolutionary and developmental biology, which could lead to the creation of a '*Homo sapiens* forma *immortalis*', as he calls it, combining cloning and stem-cell therapy (Shostak 2002). A more doubtful enterprise perhaps, though as far as I know no biologist as yet has made criticisms of it in public, nor damned it with faint praise, is that of the computer scientist Aubrey de Grey, with his engineering approach to the treatment of senescence (SENS), discussed by a professor of clinical surgery at Yale (Nuland 2005). Eschatological, nonorganic, software immortality rather than 'fleshware', embodied, immortality on the basis of speculation about the logical and physical possibilities of infinite computation has been advanced by the mathematical physicist Frank Tipler in his grandiose scheme of cyber-theology (1994), in which the questions of God, the immortality of human persons and the eternal salvation thereof, are reappraised in terms of a computational/informational metaphysic, though supposedly grounded in a reasonably plausible account of contemporary physics and extrapolations from computer science. A most ambitious work, though few will be able to follow the substantial mathematical appendices.

38 An expression coined in a treatise of political philosophy by the American Robert Nozick enthusiastically followed recently by a British philosopher (Gavaghan 2007).

39 Ectogenesis, with the entire human reproductive process being carried out outside human bodies, or indeed any natural organisms, from conception to birth, envisaged mistakenly by the biologist J. B. S. Haldane in the early 1920s as a technology for the fairly near future (unless he was writing tongue-in-cheek, not something one can discount in his case), and subsequently introduced as a major feature in Huxley's (1932) dystopia, has only become a serious prospect with the recent advances towards a viable 'artificial womb' which may be operational within the next twenty or thirty years or perhaps earlier, though premature babies have become viable at increasingly early stages so that ectogenesis in the strict sense could be seen as the endpoint of a continuum, as biomedical advances proceed along these lines (Atlan 2005). It was once proclaimed by some feminists as a key technological means for the emancipation of women, specifically from the critical burden of reproductive labour with all its social, cultural and psychological implications, though not everyone would have subscribed to the Marxist-sounding formula of female emancipation from reproductive labour, or female emancipation from the realm of biological necessity ('women in labour keep Capital in power': a slogan I once saw painted on the walls of Nuffield College Oxford), but more recently it has been perceived as ambiguous from a critical feminist viewpoint (the slogan cited was, to be sure, a pun that could be read also and perhaps primarily as: 'women in Labour [i.e., women in, supporting or voting for the British Labour Party] keep Capital in power'). Unlike men, they would thereby transcend the realm of necessity, at least partially, still within capitalism. As we noted, some authors treat ectogenesis as a continuum, with ever greater stretches of time for the gestation process taking place outside natural wombs, rather than the strict sense of the entire process, starting with extracorporeal conception till birth, taking place outside human or even non-human organic bodies (though animal bodies are increasingly exploited for the production of substances of medical value for humans, and animal wombs can also be tempting for other biomedical purposes).

40 By 'liberal' he means not authoritarian, fascist, communist or otherwise state-driven. He is less concerned with the dangers of the conditions of anarchy, failed states and civil war. It is refreshing that he did not call it 'democratic' and indeed democratic social choice mechanisms (elections, referenda, consultations, science courts, etc.) are not supposed to enter into the process of the evaluation or implementation of the new eugenics (no longer disguised as something else, being essentially re-legitimated). The term *liberal* suggests the vindication of permissiveness as against the taboos of evaporating traditional morality, residual religious objections to NRTs and the like, and vague aversive feelings (the 'yuk' factor), but some presumably 'liberal' bioethicists like John Harris have been arguing not only for the licitness, but also, more strongly, for the morally binding, the *mandatory* character of practices like human therapeutic cloning, regardless of the existence or possible discovery of alternative sources of stem cells for the invoked therapeutic purposes (as yet incipient), and, by parity of reasoning, other such human biotech techniques. Harris just stops short of claiming that human reproductive cloning is also *mandatory*, but argues that it is at least perfectly sound on moral grounds, any objections being solely the result of moral confusion, prejudice or irrationality. I am not sure whether he will not be arguing that it should be mandatory or at least encouraged as very desirable, before very long. The typical moves of much secular bioethics have been to license new biomedical

and biotech practices, to remove objections to their implementation, and to reassure the public that such practices are morally sound and should not provoke disquiet; this kind of bioethics is a new opium of the people, at least for those without religious affiliations. Now a new phase in bioethics appears to have opened up, whereby new biomedical or biotechnological practices are not only deemed sound in principle, and thus permissible, but *mandatory,* in a very striking upward shift in the force of their typical recommendations (which have generally run in the following manner: 'Accept! Accept!', even if this endorsement is generally followed by minor qualifications and cautions). The possible relevance of this new, imperious, deontic stance of bioethics (or rather of an important strand of bioethics) to the questions addressed in the paper seem too obvious to be spelled out. For they imply that conceivable new technologies in these fields may be seen not just as licit, permissible or meritorious, at least on balance, and with certain safeguards in place, but even as morally obligatory, thus entailing a *new moral burden* upon mankind, more specifically upon the enlightened vanguard, all too conscious of the new reproductive and genetic technologies, among others. This new moral burden comes at a time when humankind had been liberating itself from the repressive demands of superegos (though some critics saw this as 'repressive desublimation'), and moving towards a 'morality without obligations or sanctions' (as a nineteenth-century ethicist put it), or that in which 'it is forbidden to forbid', eventuating in the Republic of Choice. Perhaps we can all it the ultra-modern person's Moral Burden in the age of enhancement technologies. But note that eugenic movements from the beginning, in attempting to change reproductive practices to eliminate the unfit and to promote the better-quality humans, also adverted that, in order for these necessary changes for the breeding of superior humans, a strong sense of duty and sacrifice would be called for, and perhaps even a new secular religion, at least for the elites, no longer a Religion of Humanity but a religion directed towards the future, and towards higher humans still-to-come, or still to be brought into being, rather than the humans of the past and the present, at least in their biopsychical nature (as indicated by Galton, for example, and exemplified by the psychometrician R. B. Cattell's 'Beyondism', or the biologist Julian Huxley's 'evolutionary humanism', or the philosopher W. P. Montague's 'Promethean religion', and of course there were other versions) (Galton 1869; Darwin 1926).

41 Gary Becker (1930–2014) was an American economist who won the Nobel Prize in Economics in 1992. Becker theorized that marriage operates as a market.

42 E.g., 'aristogenesis', the 'beyondism' of the psychometrician and personality theorist Raymond B. Cattell.

43 Newton himself entertained the speculation that humans could disappear perhaps through some earthly or cosmic catastrophe, in accordance, of course, with divine design, and be succeeded by another human-type species, perhaps more than once.

44 Better known for his advocacy of 'situation ethics': do whatever is appropriate in a given situation, even if it contradicts conventional moral rules, according to the law of love. It was intended as a middle way between legalistic and antinomian ethics.

45 The term 'animatism', less well known than that of animism, was invented by the British anthropologist R. R. Marett, to emphasize the sense of movement of such spirit-like beings, to which perhaps corresponds to the information flow, or the property of being carriers of information ('infons') that our world view tends to impute to all beings, perhaps all matter, at least potentially (*informational animatism* we might call it in its widest cosmic sense) (Stonier 1997).

46 It is true that some other scientists have sharply disagreed with him on this score.

47 Averroism is a thirteenth-century school of Medieval philosophy based on the works of Averroës, or Ibn Rushd, a twelfth-century Islamic philosopher from Córdoba, Spain. Averroism is centered on Averroës's interpretations of Aristotle, which often went against Christian doctrine.

48 Max Ferdinand Scheler (1874–1928) was a German philosopher who produced groundbreaking work in phenomenology.

49 David Émile Durkheim (1858–1917) was a French sociologist whose influential text, *The Division of Labor in Society* (1893), introduced the concept of 'collective conscience'.

50 The formulation is cautious, because the idea certainly was around, especially in connexion with the undertakings in the organization of international scientific communication due to Nobel laureate, and chemist Ostwald, and his followers, already before World War I. Thus, the notion of the World Brain emerged in connexion precisely with the problems of maximizing the free flow of scientific communication and exchange throughout the world, and the concerns over world access and conservation of information emanating from scientific research. The idea of the World Mind stemmed from Arab or Muslim Aristotelianism, and as it appeared to negate the concept of the human person, it was deemed heretical. And when writers spoke of telegraph lines and cables as the nervous system of the Earth, already in the first half of the nineteenth century, they were close to the notion through the fascination with technical achievements in communication, especially the electric telegraph, with which human communication became independent of the transport of people, and faster than any technical locomotion facility, for the first time in history.

51 The noun had gone out of circulation since his time, although the adjective *superorganic* remained in use in American cultural anthropology until fairly recently.

52 As Pierre Musso (1987), in particular, has argued, the conception of networks in social thought is not the result of the world of the new information and communication technologies from the 1960s onwards. It emerges with the Saint-Simonians, who envisaged the entire planet as covered by networks of transport and communication, of the worldwide electric telegraph, railways, transoceanic canals, fast ships, etc., enabling the flows of free trade and commerce to encompass the globe, and the exploitation of all resources of the Earth. Their vision of the Earth as a huge workshop or factory, a factory of factories, as productive activities, and free trade, pervaded the globe, or the networked planet, has come to pass, at least something resembling it; of course today the firm, the factory, the economic organization, is seen as itself a network or network of networks, less a physical place or huge plant, with outsourcing and offshoring (Heylighen 2007).

53 For a well-known example of his development of the concept of noosphere, see Pierre Teilhard de Chardin's book, *The Phenomenon of Man* (1955).

54 *Totum* is a Latin term that means 'all' or 'whole'.

55 The semantic history of this term in European intellectual life deserves a full-fledged scholarly monograph which should take into account its considerable importance in Russia, already in the late nineteenth century, in evolutionary biology and speculative thought by scientists and philosophers, in part owing to the opposition to the emphasis on the ubiquity and primacy of struggle and competition, especially intra-specific competition, and the struggle between *individuals*, and indeed *conspecific* individuals, in Darwinism, and in recent neo-Darwinism, and on the competition, the permanent war to the death of nations and races in social Darwinism. Whereas the history of holistic modes of thought in Germany has received considerable attention, the history of holism/organicism/ symbioticism in Russian thought, important in the elaboration of the concept of the biosphere by the geologist V. Vernadsky (even if the term had been coined before him), has not received comparable attention in the West. There are many styles of holism, as there are many variants of moral and political individualism, and the cultural tonalities, the metaphysical pathoi, of holism can vary widely too, from Nazi racist holism, to the corporatist Catholic holism of Spain in Austria, the symbioticist holism of Russia, or the holism of the emergent evolution theorists, or the holism identified by Louis Dumont in India and in Germany. In recent years, the term 'symbiosi' has been used by influential biologists such as Lynn Margulis, with some impact on evolutionary biology, or Joël de Rosnay – the latter, though, affiliated to the cybernetic tradition where talk about symbiosis really got started with Licklider. The 'metaphysical pathos' of the respective uses being rather discrepant, has been widely followed. 'Symbiosis', in these texts, tends to be seen primarily as

mutualistic symbiosis, to the advantage of all the symbionts, or so it is implied quite often. It is interesting to find this valorization of symbiosis at a time of great emphasis on intensified, continuous and globalized competitiveness (world labour arbitrage) and institutionalized individualism in the economy and work life, with zero contracts increasingly appearing as the default contractual relationship, as if somehow they could be conjoined comfortably. In Russian thought, where the emphasis on symbiosis in biology, ecology and human geography flourished in the period between the 1880s and 1917, even if not entirely dominant, it presented itself as a counter-movement to Darwinism and to Malthusian population theory, as well as to some extent to Western individualistic civilization and competition or struggle, ubiquitous and unrelenting, as the sole motor of progress.

56 A metaphysical stance of brain–mind identity, in one version or another (token identity or type identity as forms of supervenience, where 'type' and 'token' are used in Peirce's sense), has become practically taken for granted in these circles as well among neuroscientists, contrary to the situation even as recently as a decade or so ago. 'Emergent materialism', 'neutral monism' *à la* Mach, or even neo-Cartesian dualism had distinguished followers in the neurosciences; Roger Sperry, a neuroscientist of Comissurectomy fame, and Sir John Eccles, another brain scientist, for example, subscribed to the former, or to some version of the autonomy of the mental or consciousness. However, this kind of view has been strongly held in medical and psychiatric circles since the middle of the nineteenth century, to go no further back. The main difference today lies perhaps in the widely held postulate of what one might call *global ontological informationalism*, that the basic stuff of the universe may be said to consist in information, or what one might call 'infomorphic' processes, if not 'infons', more fittingly than matter or energy, which underlies, or is conjointly held with, the counterpart of *generalized methodological informationalism* (Stonier 1990). According to the latter, the basic scientific heuristic consists in viewing all natural and human processes as varieties of information-processing, and information-related or infomorphic activities. This has served as the matrix of vigorous research programmes in practically every major field of scientific endeavour today, often called 'cognitivist', and the practical maxim of *fundamental Archimedean informationalism*, information as the paramount world lever, especially digital information technology as the motor of science, technology, economy, changing the inner world of mind/memes, brain/neurons and genes/information particularly. The use of the term informationalism in this context is not directly related to that of Castells, who uses it to characterize contemporary societies in which the ITCs play an ever-increasing role and shape new forms of life as well as the mode of production.

57 However, not all these projects are equally favoured by the enemies of the human status quo. For instance, the academic biologist S. Shostak, who has outlined at some length a fairly technical programme of research on immortalization, refers to what he calls the 'human-machine lobby' committed to cyborgification as the royal road to beyond-the-human, but suggests that the 'growth of an immortality lobby will ultimately overwhelm resistance' (2002, 197). Also noteworthy here is the publicity generated by Anthony de Grey's SENS research programme aiming to overcome human ageing, which is targeted as a preventable disease, within the next two decades, along several research tracks, even though this has received very mixed informal notices from academic biologists, in part perhaps because de Grey is a self-taught biologist coming from computer science, and that will not do (Nuland 2005). Shostak also anticipates in the same work that there will be a good market for 'immortality services' (as he puts it), just as there has been for the new reproductive technologies (*in vitro* insemination by partner or anonymous donor, proxy gestation, post-menopausal pregnancies, etc., while we wait for the coming of ectogenesis or the artificial womb, completely outsourcing human reproduction), though perhaps an even wider market, for the NRT-assisted are still a tiny, if steadily growing, minority in the world. Sex pre-determination of progeny by parental selection of the appropriate embryos will probably be legalized and widely practised in the UK in the coming decades, as appears from reports issued in August 2005. This is yet another biotech

'service' to the populace out of the genetic cornucopia, given that one could also in the same fashion require other qualities than sex, such as eye colour, potential height, native intelligence, and so forth. One is certainly on the slippery slope to overall genetic choice of progeny, though still somewhat behind the full expectations of reprogenetics proper. In any case, it is interesting to find a militant of the neo-human using the language of 'lobbies', lobbies for the neo-human and for the trans-human, which makes one realize even more acutely than before, that even though there are strong animal welfare/defence/rights lobbies, or lobbies for the defence of various endangered non-human animal species, there are no lobbies for the *humanum*, for the conservation of the human species, under the constant pressure from post-human projects of a far-reaching scope and depth of species-transformative and species-surpassing or species-replacing intent. Some of the implications of immortalization drawn out by Shostak concerning sexuality are rather reminiscent of the sexual utopias of Herbert Marcuse and Norman O. Brown, among the theorists of sexual radicalism of the 1960s, particularly of the former, with the termination of allegedly repressive genital sexuality and its existential concomitants, which have played such a large part in human life, and the exclusivity of the 'polymorphous perverse' sexual being as it appears in the English translations of Freud (the expression will have to be replaced, obviously). As for non-genital, non-reproductive sexuality, in a kind of perpetual childhood (as it used to be, perhaps, given the contemporary trends in sexualization of childhood), perhaps its time has come at last, not via cultural change, or the triumph of the sexual radicals of the sixties, or the merger of the late Freud with the young Marx, but via techno-biology. In any case prophets of immortalization, or at least radical extensions of the human lifespan, well beyond the Biblical limit, through science and technology, from Descartes to the present, rarely draw out so clearly such implications, which seem to reduce, rather than augment, choices, and the diversity of ways of life, though 'choice' (the expansion of choice wherever there have been givens, or binding ties, or filial piety) is a master symbol of legitimation for technoscientific as well as other endeavours in the West today ('free to choose', or, as a critic pointed out, 'free to lose'). To be sure, in the past such matters were protected by a certain prudery, and even radicals such as the philosophical anarchist Godwin or Condorcet, or marginalized thinkers like Comte, did not dwell too much on them.

58 *Ex officio* is a Latin phrase that means 'out of duty', referring to something that is due to one's office or position.

59 The pervasive obsession with 'performance' in microeconomic terms, in the polity, in society and not least in personal or sexual life in the last decade or so, as in the concern with 'benchmarking' in firms and industries, with records in sports, the whole thrust closely associated also with the permanent demand for the 'enhancement' of body, physical appearance, genes, intellect and every kind of capacity (*except* moral and affective ones, such as compassion, sympathy, benevolence, reverence for life or biophilia), seems to partially vindicate retrospectively Marcuse's expression 'the performance principle' as a master principle of contemporary life, if not quite in his sense. Recently, though, the proposal for a 'genetics of virtue' has been advanced, that is, the notion that genetic technology may be used to promote a virtuous or at least a more-virtuous generation, so that the Republic of Genetic Choice might lead on, surprisingly, not to a Republic of Selfish and Very Selfish Genes, but to a Republic of Genetic Virtue. Commendable as it may appear, surely this is only one of the end states that one may envisage, and not necessarily the most realistic (and actual intervention to make this happen goes against the prevalent norm of letting the market decide who gets which genes). One may recall that among geneticists who were the most vocal in the promotion of positive eugenics, H. J. Muller always insisted on the necessity of promoting compassion, benevolence, altruism, as well as intelligence (a point of view rejected by most other eugenicists). Altruism in the common-sense meaning, disinterested sacrifice for others, kin or non-kin, compatriots or not, friends or strangers, not in that promoted by the biologist W. D. Hamilton, of kin altruism, now standard and exclusive in evolutionary biology: sacrifice, yes, but only for those who share the

most genes with one (the degree of sacrifice to be incurred will be measured accordingly), so that the greatest possible number of one's genes will be perpetuated, and for that reason only. It is curious how a word invented by a sociologist, Auguste Comte, taken up especially in the sixties by Sorokin and latterly by others influenced by his teaching in their concerns with 'creative altruism' in contemporary life, has now been taken over and given a completely new and ethically impoverished sense by biologists, a sense that has becoming quite prevalent, even if it has not yet wholly displaced the agapistic one (disinterested sacrifice for others, for strangers, for non-kin). On the occasion of the movement of solidarity with the victims of the tsunami in East Asia some years ago, a psychologist postulated the existence of a gene for (non-kin) altruism to account for such a phenomenon and others like it!

60 If one can enhance, one can also downgrade, *ceteris paribus,* in technological terms; now there is a symmetry of technoscientific potency that is rarely adverted! '*Enhancement technologies*' of phenotypes, genes/alleles, genome, germ cells, the brain/mind, of performance levels in sport, work, war and other areas of life (sex, gaming), are all the rage, in the quest for the perfect body, the perfect baby, the perfect copy, not by chance but through design, not so much accomplished as engineered according to preset procedures, and the like, but if the symmetry maxim just enunciated is valid, one should beware of the reverse potential, the potential for *diminishment technologies*, to produce suitably downgraded humans for a variety of practical purposes. The bioethicist Joseph Fletcher's candid advocacy of the biotechnological/genetic production, the design of *inferior humans,* or auxiliary para-humans, to serve the needs and wants of the superior humans provides a startling example of the path that could be followed, for he simply extrapolated the potentialities of emerging biotechnologies at the time he wrote in the 1960s. At the time, he was a professor of Theology at Harvard University, but subsequently, when he became an atheist, he subscribed with equal, if not greater enthusiasm, to the goodness, perhaps the sanctity, certainly the necessity or inevitability of radical human transformation via biotechnology and genetic engineering. Quite apart, analytically speaking, from the downgrading effect that the technological enhancement of some provokes socially, psychologically, culturally on the not yet enhanced in commensurate measure in the relevant areas (what used to be called 'relative deprivation', but more recently phrased in terms of the concepts of 'positional goods' or negative externalities). The technologies of biological and specifically neurological engineering in question could presumably be deployed for both ends, for enhancement or perfecting as it is almost exclusively advertised, but also for the manufacture of truncated, specialized, versions of human beings or human-animal or human-machine chimeras. Artificial Zoology, too, is definitely conceivable, and Artificial Ethology, already a recognized field of scientific inquiry, either as a study of virtual reality zoomorphic beings or as the design of ground-level, robotic, performance-equivalents of various animals; incidentally, for Shostak, the production of immortal living beings – such a contradiction! – will start with animals, very likely with sheep, a guess presumably inspired by the notorious creation of Dolly, the most famous sheep in world history. The symmetry of these biotech endeavours is rarely considered from an analytical point of view, but the Harvard professor, one of the earliest bioethicists to acknowledge the far-reaching import of the new reproductive technologies, was perfectly prepared, thirty years ago, to entertain a scenario in which downgraded humans, sub-humans, or what he called '*para-humans*', 'chimeras' of a kind, would be produced to order through these technologies, to serve a variety of economic purposes, a proposal which would seem consistent with some critical tenets of bioethical casuistry, and the one thing he was not accused of in reply was lack of consistency (Fletcher 1974). Why is this symmetry principle of *modes of dehumanization* (as we might call it), or in a more neutral fashion, *modes of alterhumanization*, 'upwards' and 'downwards' in a scale of physical and mental capacities with *Homo sapiens sapiens* as the benchmark, not taken more seriously at a time when symmetry principles, or at least their simulacra or soft versions, are so fashionable in the social studies of science and other areas? It can be argued that there is no market demand at present for bioengineered self-downgrading

services certainly of a genetic kind. But self-mutilation in phenotypical terms can be not only legal, but sought after and even fashionable in some affluent circles, as in the notorious recent case of the demand for (literally) forked tongues as a fashion accessory and sex aid, in New York and elsewhere, and the pursuit of more zoomorphic phenotypes involving frequent and painful operations. Perhaps for procuring engineered subhumans or chimeric part-humans other than pure robots, only for perfecting or enhancement or performance-improving technologies, but one cannot foresee all the vagaries of market demand (or indeed the oddities of technology-push developments as opposed to market pull). It can be argued also that in a democratic polity and society such processes could not occur, or would be discontinued, or prohibited, though legal restraints on the market and on the widest freedom of choice of consumer goods and services, especially of a technologically intensive kind, are normally under attack, given the prevalent ideology of the Sovereign Individual, the Absolute Individual (today a libertarian concept though the historian Frederick Maitland long ago observed of the Western European despotisms before 1914, that the Absolute Individual and the Absolute State were complements), and the strong animus against any legislation or the public policies or indeed any public discourse redolent of paternalism (a powerful epithet, a curious vestige of old times, for maternalism, or 'maternalist feminism', which is rarely used in this context, could convey equally the same hints of protectiveness, tender loving care, even disrespect for others' autonomy, if need be, allegedly for their own good, but of course without the same negative charge, even with a more or less benevolent aura perhaps), even towards very young children; we certainly don't have a Spencer-type libertarian attitude towards young children in the sense of letting them find out for themselves that fire burns. Possibly, a recent development by American law professors of a current of thought self-described as 'libertarian paternalism' (Sunstein and Thaler 2003), a seeming oxymoron, of course, which notion they are at pains to rebut, may signal a weakening of this animus, which has pervaded the moral climate of public policy formation and general social attitudes in the West, though still a minority position amongst libertarians, thus resuming the sense of the importance of the 'protection of the inept' analysed long ago by the sociologist W. J. Goode, not least in the connexion with the major transformations in the systems of social security stemming in part from the changing age profile in the populations of most Western societies, and the state's declaration of its incapacity to address these matters in the classical welfare state fashion any longer. It could also be argued that the classical dystopia *Brave New World* discredited forever such a vision, and one need not be too concerned with such things outside horror movies, though the film *Matrix*, while seeming to have engaged the attention of many sociologists, it did not do so, it should be said, to judge from conversations, *qua* sociologists, unfortunately, that is it did *not* engage their distinctive sociological imagination; 'science fiction!': maybe, but as a scientist observed decades ago, we live in the science fiction of yesterday. Still, a sense of unease regarding these matters would be salutary and of course none of the foregoing considerations addresses *normative* questions properly so-called. In a more mundane setting, away from the more sinister scenarios, the race for biotech-engineered enhancements of many kinds (turning chance to choice) reduces many to social inferiority, at least temporarily, and this quest is very much part of the infernal spiral of the quest for 'positional goods' that marks contemporary economies and societies in an epoch of technological acceleration. Not just acceleration but as a notable inventor, and leading post-human prophet, Ray Kurzweil, argues, of 'the acceleration of acceleration', and therefore of more-than-exponential, hyperbolic growth of knowledge and technical capacities of information-processing and storage, for example; this a kind of generalization of Moore's Law. Indeed, it suffices that the media exposure of mostly natural-born (if partly artificial too) female and male models lowers the self-esteem of the majority of women and increasingly of men as well, and not just in the West, at least according to a number of surveys. The new *treadmill of genetic improvement*, as we might call it, that is likely to emerge, added to the 'treadmill of production' in capitalist economies, would radicalize the pursuit of positional goods, which had been tempered by the universality of provision of basic goods by the social-democratic welfare states, or what still remains of them (Gould, Pellow and Schnaiberg 2008).

61 The collapse of the undoubtedly well-intentioned Human Genetic Diversity Project shows there is no actual Party of Humanity in defence of the preservation of the generic identity of *homo sapiens sapiens*. Clearly the hour is late, in terms of the famous clock on the cover of the journal, *Bulletin of the Atomic Scientists* – which is still being published, although it originated as a response to the Cold War, and its ever-present threat of nuclear warfare – it would be a couple of minutes to midnight, if that much. The defence of ethnic and ethnically grounded cultural identity can mobilize enough effective opposition in defence of the preservation of sections of humanity, of its boundaries vis-à-vis other sections and especially the politically, economically and culturally dominant ones. It is more plausible to speak of a Party of Gaia, as in deep ecology or those who hold some version of biospheric egalitarianism, or of a Party of Animals, as well as of many identitarian parties, than of a Party of Humanity in the sense indicated. The world peace movement in its demands for general or at least unilateral nuclear disarmament could have been said to have operated as a Party of Humanity in its aim to reduce or eliminate the threat of nuclear war, which at the time could indeed have come close to destroying humankind. The threat of 'exterminism', as E. P. Thompson called it, has moved on, and has assumed other guises, not, in the proximate future, the physical extermination of humans, strictly speaking, but of a transformation that would radically alter the relations of filiation between the generations, the wholeness of Humanity, and lead on to various modes of alterhumanization.

62 And not least a super-warrior, defence looms large in the above-mentioned report of the National Science Foundation and the Department of Commerce, one of the 'enhancements' and 'upgrades' of the prospective superior humans being the ability to function perfectly without sleep for many days (this would come in use in connexion with the torture that some political polices specialized in), something only dreamt of by military commanders since time immemorial, not to mention almost complete invulnerability in war situations. In some ways a variant of the Futurist steel-rich man, always a male, though that can surely be remedied (they envisioned more metallic, and concomitantly less organic, humans for the new age, in part certainly because of their apotheosis of war in general, and industrialized war in particular), or Ernst Jünger's mechanical or armoured soldier, not only with 'nerves of steel' but with a steel-rich body, formulated perhaps independently of the Italian Futurists, though more comprehensive soldier cyborgs are under active consideration in the United States at least. The linking of metal and flesh, the enrichment of the flesh and the human body with mechanical metallic equipment has in fact been a theme if not an obsession of vanguard artistic movements for the last hundred years or so, the cinema, and meta-political projects, both Soviet and Nazi (Dyens 2001; Hellebust 2003; Theweleit 1987).

63 *Mutatis mutandis* is a Latin phrase that means 'things being changed that have to be changed'.

64 This stance was the very core of what we can now see was one of the most acute prophetic tracts of the twentieth century to be written by a scientist, really a polymath, John Desmond Bernal's extraordinary 1929 pamphlet *The world, the flesh and the devil*, which sketches a path of systemic transformation of the human species to a post-human future in which the successors to *Homo sapiens sapiens* engage exclusively and unrelentingly in scientific-technological research, scientific inquiry in this vision always involving engineering realizations, a quest throughout the universe till the end of time. Although a residuum of corporeal humans would be left on earth, incorrigibly addicted to the pleasures of the flesh, the table and the bed: at any rate, the human vanguard now turned into mechanized post-organic, digital beings, would have left. The paramount conviction of the author seemed to be that the point of human beings, in the last analysis (or to put it in more traditional terms the answer to the question 'what we are here on Earth for?', or if you like, our species-*Beruf*), is the pursuit of scientific knowledge to the utmost extent, and therefore when the biological limitations of humans stand in the way

of this pursuit, dehumanization or transhumanization should be instituted by some humans to let intelligence develop through other vehicles than the human body, or indeed any organic body, a merely provisional medium for this pursuit, now outdated. In recent decades, this extraordinary visionary work (the vision was never abandoned by the author but also never elaborated in the rest of his active life), has increasingly gained topicality and relevance to the post-humanist projects of our time, although written long before the computational, cybernetic and molecular–biological revolutions. However, Bernal (1939) did promote work in molecular biology in his laboratory, known worldwide for its exceptional gender equality and non-hierarchical relations, a laboratory where Rosalind Franklin, a key figure in the story leading to the discovery of the double-helix model of DNA, was able to work, and was a pioneer of studies in scientific communication and documentation.

65 Given the long history of the word 'artificial', it is apposite to recall just two examples: that Hobbes wrote of states or Leviathans as Artificial Persons, and according to the US Supreme Court in an epoch-making decision, business corporations, legal artefacts, were just like singular natural persons, insofar as that could help their profit-making activities and diminish their fiduciary responsibilities. A provision that has stood the test of time, despite the anti-trust movement and the various surges of federal regulation of business corporations and recurrent episodes of trust-busting, as in the breakup of the Bell Telephone Company, whose laboratories did power the cybernetic revolution as well as much science of a fairly pure kind.

66 This includes some of the higher primates, as in the Great Ape project, possibly also dolphins, as has been argued at considerable length by a philosopher, potentially some robots, as argued in a report by the Principal Scientific Officer to the UK Government, Sir David King, who asked us to be prepared to recognize the civil, legal and even the political rights of the intelligent robots in the next few decades. He did not, I think, anticipate conjugality and sex, or even mutual affection, between humans and robots. Of course, some, though not all, the issues of robot ethics and jurisprudence were anticipated decades ago by Isaac Asimov's 'laws of robotics', which have been ripe for extension for some time.

These may comprise foetuses, the incapacitated aged, the severely handicapped, children with a number of genetic defects, and more. They overlap, one cannot fail to notice, with some of the key target categories of negative eugenics from its inception, which also aimed to euthanize such people or prevent their procreation in due course. The disfranchisement of these categories of persons or putative persons is quite striking, given the expansion of the status of personhood, traditionally practically coextensive with the realm of humans, advocated by the same philosophers to nonhuman and even non-alive or anthropogenic categories of beings, objects, entities. Sometimes the argument is put as follows, not least by Minsky: they are not like us, yes, but they are 'our mind's children', though perhaps he meant something like 'they are our mind's children, the finest children of the finest minds, and, like the finest human children, they will surpass us, succeed us, replace us, when the time comes for us to die, naturally or by the euthanasia we will be well advised to solicit, or which will be kindly administered by them as we will be cosmically superfluous as individuals or as a species'.

67 His last book was entitled *God and Golem.*

68 Even if to the disgust of such salient and tireless critics of human 'speciesism', 'species chauvinism' or 'meat chauvinism', as Moravec and Minsky like to put it, as despisers of the biological brain as but 'meatware' or 'fleshware', and of the organic body in general compared to life *in silico,* to electronic calculating machinery undergoing accelerating progress in terms of years and not biological generations let alone evolutionary time. This is expected to shortly catch up with and overtake natural human intelligence, even if this has been rising steadily over the last forty years or so in Western countries and Japan, according to the thesis of the so-called 'Flynn effect' (Flynn 2007). The result will be to generate the greatest 'intelligence explosion' growth of information-processing, calculation and computational capacities in general, and information explosion on earth, propelled by Moore's Law and allied trends, intelligence here

not meaning so much enhanced human intelligence via neurogenetic or brain-enrichment means, as non-biological, artificial intelligence with the super-exponential, hyperbolic (in the mathematical sense, but it will do in the rhetorical sense as well). And not just on Earth, in the solar system, perhaps even in our galaxy, if not indeed in the cosmos. Numerous estimates have appeared of the total information stored in our computer memories and other media such as libraries, of the aggregate information stock on Earth, and sometimes even of the total amount of information stored by living beings or the totality of biospecies on Earth. Among the most vocal exponents of this outlook can be found in computer and robotic scientists such as Minsky and Moravec, who, in addition to anti-humanism, profess atheism, reject the ontological autonomy of the mind, the soul, and anything remotely spiritual, ruled out by their strict physicalistic ontology. Their strict physicalism, as that of many other computer scientists, excludes positions like that of emergent materialism (in which mental phenomena depend on their neural base but cannot be explained or predicted entirely on the basis of the state description of the brain), which still has its defenders, and it is not clear whether they would concur with the well-known 'anomalous monism' of Davidson, according to which, while brain-states and mental states are identical no strict laws connecting the two can be found. Strict laws in the sense of enabling one to predict mental phenomena, at any rate 'propositional attitudes', often summarized as beliefs and desires, from knowledge of brain states, not to mention the vexed problem in the philosophy of mind of the intractability of qualia. I fear even anomalous monism would be too much of a concession to ghostliness. As to panpsychism, which enjoys a considerable revival today in quite sophisticated ways, and it is not the monopoly of cranks and charlatans, and in any case, was professed by such a thinker as Spinoza (recently evoked by a neuroscientist of distinction, António Damasio) and by some distinguished physicists of our time. I imagine they would regard it as beyond the pale, and not worthy of any serious consideration; Neurath might well have placed it in his proposed Index Verborum Prohibitorum.

69 According to a 2011 article, one out of every fifty US troops in Afghanistan is a robot (Axe 2011). There are also reports of 'undue' emotional attachment of soldiers to 'their' robots, i.e. robots entrusted with their protection, as evidenced in their reactions to the destruction of the robots by enemy action or otherwise.

70 This is actually the title of an enthusiastic, annunciatory book.

71 An absolutely indispensable word, coined by the linguist Geoffrey Sampson, which surprisingly has not been widely adopted yet.

72 *Prima facie* is a Latin phrase meaning 'at first appearance'.

73 These remarks will seem bizarre to the cognoscenti, but I have added them in the light of comments made to me by distinguished natural and social scientists in a certain country. Of course, there are always intelligent and alert people who, somehow, have not heard the news, whatever the news is, no matter how long ago, and however bad. Or perhaps have forgotten it.

74 As the transhumanist movement has grown and become more international (with offshoots in various parts of the world, including Latin America, and I have no doubt will recruit supporters in Portugal if it has not done so already), and more formalized, it has naturally attracted wider public notice, and a non-technical critical work by a Christian scholar has already appeared on the subject (Smith 2004; Alexander 2003; Bailey 2005).

75 *La société sauvage* is a French phrase that means 'savage society'.

76 *Fiat scientiae* and *pereat mundus* are Latin phrases that mean 'let there be knowledge' and 'let the world perish', respectively.

77 *Mundus* is a Latin term that means 'world'.

78 *Mon semblable, mon frère, ma soeur* is a French phrase that means 'my likeness, my brother, my sister'.

79 *Unicum* is a Latin term that means 'unique' or 'one of a kind'.

80 *Tremendum* is a Latin term that means 'awe inspiring'.

81 *Umwelt* is a German term that means 'environment'.

82 'Mind uploading', the transfer of the total information in the brain of a human being to a computer that can incorporate it, and therefore of superior intelligence, is still a distant goal, not like the unrelated goal of teleportation. It would involve huge advances in neuroscience, in imaging, in computational power and Artificial Intelligence, but these are all areas in which advance is rapid and escalating. Oddly, though, a French math prodigy claims that he aims at running something like a computer programme in his mind, which he can use for any task, and then download it to a computer programme so that the computer could become mindlike (*Daily Telegraph* 2007). He calls this 'reverse artificial intelligence', and one can see why. But it is not easy to see that the catch-all term 'cyborg' quite captures this phenomenon. Perhaps he wants to be present at the creation.

According to Minsky, the computer scientist, roboticist and robot ideologue to whom we have referred more than once: 'Ordinary citizens would not know what to do with eternal life. The masses do not have any clear-cut goals or purposes'. He is paraphrased as saying that '[o]nly scientists who work on problems that might take decades to solve appreciate the need for extended life-spans' (*New Scientist* 2007, 46). This an empirical claim that has not been put to the test. In any case, one can think of many – other than scientists – who might well appreciate the need for extended lifespans. Conversely, burnt-out scientists, or scientists who achieved just one significant discovery and know they will not make another, might not be so keen on the idea, either.

Chapter Three

PERPETUAL AUGMENTATION: FROM EUGENICS TO HUMAN GENETIC CAPITALISM

Should the question of human reproductive cloning be relegated to the realm of techno-futurology if not techno-fantasy, beyond the purview of ethics, or is it amenable to some kind of serious philosophical reflection? The question has been addressed at some length by philosophers of morality, of law, of biology, of technology, as well as theologians, for some time. Already in the 1970s, thinkers of the stature of Hans Jonas had addressed the great width and depth of implications of the matter, and of course in the last few years, after the Dolly landmark in 1996, the pace of pertinent publications has accelerated, and numerous publications and collections of papers have appeared by serious scholars in a variety of disciplines in a number of countries. Human reproductive cloning (HRC) has not yet come to pass, as far as is known to the world, but there is already a corpus of thoughtful writing on the matter, as well as, to be sure, much facile – even ribald – comment from people who ought to have known better.[1] By contrast, animal cloning, reproductive or therapeutic, has received little formal philosophical attention, and the ethical or other value issues raised by human therapeutic cloning, although addressed and referred to in the press, have been, perhaps rightly, overshadowed by those of HRC.

HRC is most obviously assigned to the genus 'reproductive technologies', or 'new reproductive technologies', which has expanded considerably in range, sophistication and deployment in the last three or four decades, though it surely constitutes a distinct species within this genus, at the very least.[2] It is at least problematical that the social, ethical, axiological, metaphysical implications of this potential new species of human reproductive technologies can simply be regarded as perfectly continuous with the previous or current technologies of medically assisted reproduction (IVF) or ART, that have been implemented and accepted so far in the West. However, it should be noted that conventional *bioethics* restricts itself to ethical issues without much if any consideration of total axiological and, more broadly, 'evaluative-metaphysical' implications regarding especially the metaphysical status of the human person and the ontology of persons-in-society.[3] It is not *biophilosophy* in a comprehensive sense, which would seem a desirable project in this area as in every other; a restriction to ethics narrowly construed could be pretty constricting, just as the restriction to epistemological issues diminished the reach and interest of the philosophy of science for so long. And even as ethics, conventional Anglophone bioethics, with some notable exceptions, addresses and reflects on standard liberal Western moral values – or at any rate a special package of these as if the liberal West were its sole constituency, *de facto* and *de jure*[4] – it does not seek to take seriously, let alone to learn from, the cultural diversity within the West (and within the consciences of individual Westerners), let alone further afield, despite the acknowledged importance of descriptive (not normative) cultural relativism and the much-vaunted 'conversation of humankind'.[5] It is as if it had never heard of the Weberian 'polytheism of values', of the aporeticity of values, of the tragic sense of life, of the incommensurabilities if not incomparabilities of value systems such as a widely read exponent of contemporary political liberalism, Isaiah Berlin, constantly addressed, precisely in order to bring out that the belief in the unifiability of values was associated with authoritarianism.[6] It is indeed remarkable how little the incommensurability problematic, still a troubling issue in the philosophy of science after forty years of discussion in the wake of Hanson's and Kuhn's provocations, and much invoked elsewhere, seems not to impinge on bioethics, which might well appear as the most commensurabilist or commensurabilized of domains, as if moral language were exempt.[7]

Anglophone bioethics do not spend too much time clarifying our intuitions, and exhibiting the many-valuedness of moral situations, recalling the teachings of varied moral traditions, to bring into purview manifold factors and considerations that tend to be neglected in a technological civilization. The professional bias of conventional bioethics, or the thrust of a lot of it, seems to be, when addressing new technologies, to come to a decision, *la rage de vouloir conclure*,[8] as if it were a 'policy science'.[9] Rather, since it is not in the running for scientific status, it comes across as a kind of 'policy philosophy', paradoxical as it may seem. By 'policy philosophy' I don't mean philosophical reflection on the nature of policy, or philosophical reflection to aid policymakers, but philosophical discourse that aims primarily at producing policy recommendations, much more so than 'practical philosophy' or 'applied philosophy' as they are usually understood. Put another way, bioethics leans more towards *biopolicy* than towards the more reflective, the more searching or at least more patient, kinds of inquiry that might be called *biophilosophy* or even *biosophy*. 'Biopolitics' was coined with a different sense, otherwise one might suggest demarcating 'biopolitical' or 'biopolicy' studies from bioethical or 'biophilosophical' ones. Given the lexical and institutional entrenchment that has already occurred it is most unlikely that this will ever happen. Moreover, the upshot tends to be, more often than not, one of ratification, accompanied from time to time by the pious hope for a pause, but no more than a pause, to take a breath, given the onrush of biotechnological change. Such an inclination is as might be expected in such a technological civilization, with its strong bias towards the implementation of new technologies as they come along. HRC may now be an exception, in legislation more than in academic bioethics, and indeed such legislative prohibitions are likely to annoy many bioethicists, with a general presumption of beneficence or a potential for beneficence, as well as adequacy being attached to them *ab initio*. Conventional bioethics has lost all capacity to surprise; it rather seems like a game with preset constraints as it moves to conclude, 'Yes', with more or less emphasis, and with only slight qualifications and reservations, with regard to every biotechnological project seriously entertained and pressed by reputable scientists. I am reminded of the two-party system set up by the Brazilian military dictatorship in the 1960s – a system which, according to the bon mot of the time, consisted of the party of 'Yes!', and the party of 'Yes, Sir!'.[10] A Martian reading this literature, or much of it, would have no sense that Earthlings have been experiencing a series of moral, theological, metaphysical earthquakes owing to the novel, rapid, unsettling character of biotechnological change. To be sure, perhaps only minor ones, compared to the large-scale political and economic upheavals of the twentieth century, the genocides and ethnocides of this terrible century, but no less real for all that.

This bias for the ratification of new technologies, and therefore for the acceptance of what inevitable consequences may flow from their insertion into extant social structures, seems to apply quite fully to the stream of new genetic and genomic techniques. After all, they promise what specifically 'medical utopias' like the nineteenth-century *Hygeia,* or the biological side of the most fantastic social utopias had envisioned.[11] These include health, or rather perfect health, throughout life, not just freedom from disease, cures for all ills, over the last decades above all cures for all cancers. In the field of reproduction, this entails perfect babies, reproduction without sex, or perhaps even gestation, feasible at any time (or up to 60, 70, 80 years of age), the defeat of all constraining biological clocks.

In the field of macrobiosis, this includes longevity, perhaps even immortality, or cryogenic waiting until immortality comes along, great bodies at any age, great looks at all times, and superb intelligences in permanent tip-top condition through an ever-longer life.[12] In these cases the process includes 'regenerative medicine' or the replacement of decaying or ageing tissues in the elderly by new tissues, grown in vitro and then implanted – not simply genetic or genomic engineering. But besides these intragenerational achievements, there also beckon *transgenerational* psychosomatic advances for one's progeny, on demand, and even the indefinite genetic – and thereby somatic and cognitive – improvement of the species to average levels beyond all records, surpassing what any human being has ever reached in terms of brains or brawn, genius or muscle, without foreseeable upper limits, such as utopians like Fourier or scientific socialists like Trotsky envisioned, as well as biologists of considerable standing like Alfred Russel Wallace (the co-discoverer of natural selection) in 1864, and H. J. Muller in 1935, all in very similar terms, and apparently in complete independence.

What has receded from view is the kind of psychological utopianism in which perfect positive mental health is desiderated and the satisfactions of the flesh and psyche are achieved at all levels, self-knowledge and self-actualization, so characteristic of the nineteenth and twentieth centuries' genre that a great historian of utopias called *eupscyhias,* appears to have faded away. True, the diagnosis of depressive mental illness has reached epidemic proportions in Western societies, and the anti-depressant industry has grown and grown to meet the challenge, or at least the effective demand. There has been in recent decades an increasing tendency to impute exclusive or primary causality to genes to account for depression or mental illness in general, contributing thereby to the decline of non-chemical psychotherapy, although genetic therapy for this range of illnesses is not yet on the scale for organic ones. It was plausible in the 1960s to speak of the 'psychotherapeutic society', of the permanent and almost mandatory psychoanalysis, the 'talking cures', at least for certain sections of the population, involving not only a 'normal' resort to therapy but much more than that – a cultural climate in which psychoanalytic concepts were widely diffused and shaped common self-understanding or at least the dominant vocabulary of self-appraisal. The psychopharmacological society, envisaged by some dystopias, is coming into being with the rise in diagnosed depression and the greater emphasis on chemical medication with psychoactive drugs in its treatment, with ever-larger numbers of patients, is taking its place. However, the richness of the schemes of cultural understanding of the psychoanalytic era, with its far-flung hermeneutic resonance – although there were always tensions in the interpretation of psychoanalysis, especially in the Freudian varieties, between physicalistic or energeticist readings, and hermeneutic, symbolic, *verstehen* ones[13] – has not been emulated. In any case, the fashion of genetic aetiology, the geneticization of mental illness, is replacing psychological and sociological aetiologies, though in a way it can be seen as a resurgence of the supposedly scientific hereditarian explanations of mental illness that have flourished in the last one hundred and fifty years or so (with Lamarckian as well as Darwinian/Weismannian construals of biological inheritance). This development is in association with eugenics, with new and purportedly superior, non-Lamarckian, molecular-biological, truly scientific grounding, and with the added promise that genetic intervention – the injection of the appropriate alleles in the fertilized ovum – will prevent depression in future generations. Social causality is also thereby discounted or relegated to a very subordinate place in models of causal explanation of mental illness, although recent findings, such as the increase in suicide rates under right-wing governments in several Western democracies in the last three decades, detected independently by several researchers who have applied statistical analyses to the official data, would be hard to account for under this kind of genetic aetiology. In any case if the volume of depression outcomes in any given society, so to speak, is more or less constant, or may be presumed to be so (constancy being as it were the base-line), unless there is strong evidence to the contrary, the steady, substantial, unrelenting increase in the numbers of persons so diagnosed throughout the Western world, which shows no signs of levelling off, cannot presumably be due to genetic factors per se. Perhaps the proliferating happiness manuals, the burgeoning psychological studies of happiness fill the void, and the Internet and the new 'e-topias' take the place of the eupsychian dreams, though, as happens in other spheres, the *right* to happiness (a misconstrual of the Jeffersonian right to the *pursuit* of happiness) can turn into a vexing obligation, the *duty* to be happy.[14]

Slippery Slopes and the Appraisal of Reproductive Technologies

HRC need not be addressed afresh in any depth now or ever, certainly not after the *fait accompli*,[15] as we have learnt from previous cases of technological implementation. This is because, according to conventional bioethics (a) we already know basically all there is to know, at least in terms of moral knowledge about the new human reproductive technologies; (b) we already know they are perfectly okay, by and large (the qualifications are matters of detail and circumstance), in terms of moral appraisal – in other words, we have already come to a verdict, and that's final; and (c) HRC is just another of these, there being nothing *sui generis* about it,[16] in any ethically relevant and consequential particular.

Now claim (c), it seems to me, is question-begging, because while there certainly are obvious positive analogies between HRC and the older reproductive technologies

(RTs), equally obviously there are some dissimilarities – crucially the asexuality of ARC. Whether this and all the implications and reverberations in society, culture and the psyche, that might spring therefrom (depending partly on how widespread the practice might become and, indeed, the potential for so spreading), makes essential or at least significant differences in the total axiological situation, is precisely what is at issue. Arguably, the feature of asexuality or biological monoparentality (social monoparentality has long preceded it), is a novelty of far greater order involving a biotechnological intervention of far greater import than any of those involved in the human RTs that have become familiar in the last thirty-five years or so. Those who object will undoubtedly regard the perfect continuity thesis as a prime example of the slippery slope. It is true that the slippery-slope argument (SSA) has been used regularly against new technologies, so much so that its use has been assessed in this context almost exclusively.[17] But it is less-often noticed that advocates of new technologies, and most certainly advocates of human RTs, regularly appeal to the argument that if a new or prospective technology is continuous with an existing, already-accepted one, then it should certainly be adopted, which is of course the positive version of the SSA.

Sociologists have often shown how entrenched technologies are so taken for granted, so backgrounded, so thoroughly implicated in the fabric of our daily lives, as to become 'naturalized', unquestioned, tacitly relied on, unthought about except intermittently on occasions of breakdown – and then to little effect, if one thinks of the scale of the social movements that have sprung from the breakdown of the great 'economic machines' in great crises – and above all the Great Depression. Those who seek to legitimate technologies to come on the grounds that they are perfectly continuous with those already naturalized in this sense, or in the process thereof, seek to bring about, or to help along, a kind of anticipatory or perhaps *propitiatory* naturalization (that seems to be what much of bioethics is about). More generally, the argument seems to be that after a while, every technology, including biotechnological and genoengineering technologies, already pervading our lives, may yet reach similar proportions, sooner or later can and will be 'naturalized', at least in a progressive society. This principle applies, then, to technologies involved directly in what one might call *second-order* or *reflexive anthropogenesis*, the deliberate technological modification of the human being by humans, especially involving wide-impact, durable, inherited changes in soma and psyche – in contradistinction to the technological modification, intended or unintended, of the nonhuman environment by humans or simple anthropogenesis, regardless of area, level or scope, permanence, heritability or irreversibility. (This might be confused with a 'reflexive anthropogenesis', which would involve a critical scrutiny of ongoing human-induced technologically and economically mediated modification (anthropogenesis) of the environment, but by 'reflexive' here I wish to draw particular attention to the modification of human beings by human beings through the instrumentality of advanced technologies of biological reshaping, particularly genetic and genomic germinal engineering.)

The central dogma underlying this type of argumentation seems to be an *asymmetry thesis* that resistance (resistance of any kind) to technological innovation of any kind is what calls for explanation and research (and remediation), not acceptance per se, and indeed the classical 'cultural lag thesis' in the sociology of technology involved precisely this assumption. To some extent this applied also in the economics of technology. This has been qualified in regard to alternative technological options, as indicated by the 'QWERTY syndrome', that is, the differential acceptance and entrenchment of an inferior technology and rejection of a superior one, though the historical accuracy of the specific example of the keyboard has been questioned, that being the fate of historical examples in general. To my mind what is striking is how such technological projects have met with so little opposition given the staggering character of what they purport to bring about, compared with the ferocity of the struggles over political proposals and social inventions, and the general animus against social engineering. The increasingly widespread practice of genetic engineering, and of human bioengineering as a whole, never seems to count as social engineering, no matter how widespread, how novel, how unforeseeable its impacts, because it seems only pernicious social engineering, as seen by the commentator, counts as social engineering.[18] Could it be that the ongoing and emerging technologies of second-order or reflexive anthropogenesis will elicit feebler critical responses than any major surge of technological innovations since the early phases of the Industrial Revolution? This at a time when the evidence of global climate change and more importantly of the cumulative anthropogenic or technogenic impact on global climate change – above all since the Industrial Revolution – rejected out of hand for so long within and outside the scientific community, is finally coming across to most people.

Most of the bioethicists I have read would deny that the slippery slope obtains, on the grounds that there never is any such overwhelming momentum, irresistible thrust or drift, and hold that we could stop at any time, in a series of closely related biotechnological innovations if and when significant unwholesome effects were discerned. This seems more a matter of optimistic faith than of anything else and, as usual, it is backed by a secondary thesis that remedial technologies for the mitigation or elimination of the damage technologies inflict, often unforeseen, can always be found in good time before large-scale irreversible, or practically irreversible, harm has been done to local, regional or planetary ecosystems (to give an example of the kind of impact that may occur). According to this standpoint, for any given technology N, a subsequent N+1 remedial technology can, in principle, always be found, and in good time, too, to reverse or compensate for the ill effects that have ensued of any and every kind. Such staggering claims cannot be vindicated rationally; though arguments have a role, ultimately these are matters of faith, which in the last analysis pertain to the theodicy or eschatology of technology, not to the 'public understanding of science' or 'better science education', under which label it is no doubt hoped to 'sell' science, or rather packages of technoscience wholesale, to secure acceptance or at least resignation, and not to enable members of the public to look at science more critically or in a wider cultural and historical framework than that normally perceived by very busy, permanently grant-seeking, working scientists.

A tacit assumption in the (a) and (b) theses indicated above is the presumption that once RTs have been accepted, they are here to stay and, because we have accepted them, we can take it for granted that the question of their overall axiological configuration, of values and disvalues, rightnesses or goodnesses, need not, or must not, even be raised ever again – at least, some might add, not within what one might call 'the moral present' or the 'moral space' of the West, the tacit boundaries of conventional bioethics.

But why should we accept these boundaries? Surely, they cannot be treated tacitly but wrongly as exhausting what we might call the *total moral ecumene* of our time,[19] the field of legitimate participation in the moral conversation of humankind, the actual or potential *interlocuteurs valables* on these matters,[20] involving a diversity of cultures and the plurality of civilizations, as we live increasingly in a 'cosmopolis', or would like to think so, in a shared world genetic and epidemiological as well as a cognitive pool.[21] Perhaps those who insist or persist in questioning the validity of the acceptance of RTs or the impending HRC are deemed to be anachronics, or outliers, moral aliens (beyond the pale, like the 'creationists' for evolutionary biologists), rather than prima facie fellow-members of this moral ecumene. But it is difficult to see how acceptance of RTs can be taken as conclusive as far as normative validity is concerned (a socialized version of the naturalistic fallacy), and how acceptance of each new RT, up to and including HRC, regarded as irreversible in fact and in reason and in value, by a kind of *ratchet effect of naturalizations*. This is odd, because according to the principles of the corrigibility, revisability, falsifiability, criticizeability or, more inclusively perhaps, 'the permanent control' of all claims, whether cognitive, methodological, theoretical, evaluative, subscribed to in Popperian or other critical rationalist terms, it could not be ruled out that in the light of subsequent experience with the field of operation and other inputs, for example, of theoretical reflection with or about RTs or other emerging, broader or fuller contexts of appraisal we might come to reconsider and reassess what had been accepted, for nearly everything is corrigible and much can be contingent (in a manner analogous to the proceedings of the hypothetico-deductive method or the procedure known as the 'wide reflective equilibrium', though this leans perhaps towards excessive conservatism unlike the Popperian falsificationist version of the former).[22]

Reappraisals of the human past, not least of the recent past, in analytical, explanatory or evaluative fashion, are the very stuff of cultural modernity as well as, of course, the imperative of academic industries. There is no reason to exempt current, accepted, even entrenched, technologies from reconsideration and reappraisal, in the light of their unfolding impacts, not least ecological and human-biological, and of better knowledge and understanding of these. A few examples may be given of acclaimed and widely used technologies being sharply reduced or discontinued altogether after decades of indiscriminate application throughout the world. We have, after all, come to cut down on the use of pesticides in the light of their myriad unintended biocidal and ecocidal effects. The constant X-ray testing, especially of children, that was a current, unquestioned, even mandatory practice for decades in the richer Western countries, has been discontinued. The practice of lobotomy in human psychosurgery, so enthusiastically endorsed for some decades, was then rejected. The spread, banning and then unbanning of ECT, the widespread use of CFCs, a marvel of beneficent industrial chemistry, supposedly totally safe, deployed for several decades, has also been restrained. There is even the discontinuance of the building of new nuclear power stations in a number of countries, even in the United States, when originally, in the 1950s and 1960s, the building of thousands and thousands of such power stations covering the length and breadth of the entire country, was enthusiastically contemplated, without any serious qualms. Of course, even now there are those who demand a return to the intensive use of synthetic chemical pesticides, not least of DDT, at least in some parts of the world, and the building of new nuclear power stations if for reasons which had not been paramount in their early projects, especially to reduce the use of fossil fuels with their inputs into the greenhouse effect, and so forth.

Homo Heterarchicus

In an increasingly hyper-technological society, we are becoming the nodal points for the intersection of multiple technologies and technological systems which, through universal digitalization and informatization, are increasingly interconvertible, both amongst themselves and in and through us, despite our perhaps irredeemable partially analogue nature where, in a pluralist society, we had become the nodes for the intersection of multiple non-concentric social groups or circles.[23] Participation in a web of non-concentric, only partly overlapping, group affiliations was supposed to be a major factor in fostering the emancipation of the person. Whether location in cyberspace and in technological networks increasingly integrated may conduce to emancipation is arguable, for the person who could stand for the *homo aequalis* and partaking of varied intersecting groups for privacy and achieved moral autonomy, may not carry over as a node in universal digitalization and reticularization. Perhaps the new type of human being – the metonym of intersecting, criss-crossing, hyperlinked technological systems – will take shape as a *homo heterarchicus*, the locus of networking and the networking of networking, and so on to any degree of recursivity. The propaganda that apotheosizes the Information Age as a universal solvent of hierarchies of all kinds – much more effective and far-reaching than political revolutions of classical liberal individualism of *homo aequalis* and the techno-economic regime of mass consumption industrialism – would imply that the promotion of reticularity and omnilaterality would be optimific, and conducive to all good things. But under these circumstances, can Kantian-type moral and intellectual autonomy be reinstated, even as an ideal, or would the slogan of 'heterarchy' stand euphemistically for *heteronomy*, if no longer in the organic, hierarchical community, or the twentieth-century crowd, but in an online virtual community? (Moscovici 1981; Riesman, Glazer and Denney 1950).[24]

Our three ages of a cybernetic historiosophy would then be (1) the Age of Hierarchy (*homo hierarchicus*) of embeddedness in concentric organic and concrete communities; (2) the Age of Equality (*homo aequalis*) and the Abstract Society, moral individualism and rational autonomous peers in non-concentric social circles; and, finally, (3) the emerging Age of Permanent Inclusive Online Heterarchy (*homo heterarchicus*), with e-science, e-philosophy, cyber-sex, and so on, within its electronic quasi-concrete 'virtual communities'. To be sure the process of heterarchization is only incipient, for not everyone is online all the time with equal competence and receptivity, not every artefact has been interconnected by Bluetooth technology, not all technologies have been informatized and digitalized, and it will take time for the heterarchic mode of being to become fully 'naturalized', despite all the incantations already in process of propitiatory naturalization, indulged by enthusiastic virtualizers.

The Ratchet Effect, 'Reverse Serendipity' and Adiaphorization

But we do not, in effect, even in this kind of society, just go on extending accepted technologies and arguing from their acceptance that analogous ones, though perhaps more sophisticated or potent, are therefore to be accepted likewise. Sometimes we have gone back and rejected technologies that had become widely used and, indeed, to some extent *because* they had been widely used and used as in the case of the grotesquely massive and ever-growing utilization of pesticides in all sorts of ecosystems and agrarian systems of the three decades following the end of World War II, so that the damage had reached global proportions and even insofar as it is not strictly irreversible, calling for the utmost efforts of 'restoration ecology', involving what an economist has called 'reverse serendipity'.[25] We are surely capable of learning from experience in a wide sense of 'experience' (and that includes maturing self-reflection) and coming to see the impact of certain technologies and biomedical practices in a new light, with an appreciation of not only past and current impacts, but also on future generations, a moral category that has become more salient owing to a considerable extent to technological change, whether directly or indirectly. Though this extension of the moral horizon for technological appraisal has been questioned by some moral philosophers (e,g.: How can unspecifiable individuals have interests, or claims? Can we possibly owe any duties to posterity?), it seems well-founded.

We do not just proceed then from hitherto accepted practices such as the RTs that have become established or even naturalized though never uncontested by some currents of opinion, even in the West, in the last twenty or thirty years or so to the consequential acceptance of a new RT, to be added to the existing repertoire because previous ones have become established. This seems an astonishingly straight, direct linear inference, drawn sometimes by the very scholars who pay lip service to the non-linear 'wide reflective equilibrium' that John Rawls and, subsequently, Norman Daniels (which tend towards a kind of Coherence Theory of moral truth) have theorized.[26] What is 'wide' about this type of inference with its single input of precedence? In our individual and collective lives, we may come to see our biographical past in a new way, though some might claim that repentance, remorse, a sense of guilt, expiation and the like no longer have a place – or just a very restricted hold – in our current ethico-political scheme. The Protestant 'guilt culture' of Americans and Northern Europeans, so often counterposed to the 'shame culture' of Mediterranean countries and other areas bogged down in economic underdevelopment, has receded, except among 'born-again' Christians and in remnants of official piety. Neither guilt-ridden nor shame-faced, the new Protean Man and Protean Woman both expect and seek repeated drastic personal change, and responsibility for the past selves counts for little, at least in a non-instrumental way.[27] But the Protean Human need not be paradigmatic. In the appraisal of technologies we can also come to see the glorious recent past in a new way and not the present, as but an extension of a presumptively valuable or even self-justifying repertoire. That indeed would be to participate in and to condone what we have called the *ratchet effect of naturalizations* with no end in sight – not only for RTs of any and every kind but for the whole project of unceasing second-order anthropogenesis. But there can be and there ought to be retrospective retransmission of disvalue or invalidation, just as for the extensionalists there is a presumption of transmission of value and validity from the currently or recently accepted to the presumptively analogous emerging or projected biogenetic, genetic or 'reprogenetic' or other biotechnology or RT. This kind of argument affords a nice and easy way to deproblematize and neutralize all the putative value-laden, profanation-seeming implications of technological novelties that folk ethics, common sense moral intuitions or gut-feelings, or religions may find (as is surely the case with HRC) very disturbing.

But then technological innovation, not least with respect to human reproduction, can only proceed by constant *adiaphorization*, which might be called *strong naturalization*, inasmuch as it implies not only acceptance or taken-for-grantedness but the voiding of value-implications, or value-stakes, the neutralization of what had been charged with sacred connotations, reason enough for bitter disputes, even desacralization or denuminization of previously value-charged fields of human life and social practice. This was even strongly so in the case with modern capitalism, as Carl Schmitt pointed out in his account of successive neutralizations in the West since the Reformation. In the domains of religion, the term *adiaphoron* came into wider currency owing to those who wished to put an end to bloody religious confrontations within Western Christianity, of the economy, and of politics. Schumpeter saw the same modern capitalism, not modern science, as the destroyer of the pre-logical mentality of mystical participation delineated by Lévy-Bruhl, whom he cites, curiously, instead of Max Weber's disenchantment of the world or self-expulsion from 'the magical garden of the world' (Schumpeter 1942; Weber 1930). It was not science in modern Europe, that, initially at least, opposed, or had to oppose, natural magic, but the Protestant religion.[28] One might even say that these processes of neutralization or adiophorization of things and living beings enable technological innovation as much as calculation and experimentation, though these processes of neutralization can be partially reversed, as the example of animal rights protests and the shift in moral sensibility towards animal pain and suffering of the last two decades shows. This is not an inconsequential matter taking into account not only barbaric farming practices, which still continue on a large scale in much of the world or the biotechnological production of such vaunted new animal varieties as featherless chickens, or also the yearly use and disposal of hundreds of thousands of laboratory animals.[29] In a more complex way, consider also the ongoing disputes over 'genetically modified organisms' or 'genetically manipulated organisms' (GMOs), whose deployment is being pushed by the government in the UK *against* public opinion. It is not just a case of 'technology-push' or 'corporation-push' (or 'market-pull'), but also, and perhaps critically, at least in this case, of 'state-push', paving the way to their application, protected by the police, the security services and the courts; it is not just nature that has to be overcome and 'domesticated' by new technologies, by new 'scientific miracles', but the people, the unenlightened people, or the new *classes dangereuses* (they have no right to be 'Luddites'!). Every species of organisms can be improved, and we already have or will have featherless chickens (although chickens in factory farms still have to be debeaked non-genetically), super-tomatoes, 'super-onions' that will not make your eyes water, and so forth.

Before too long we will have to talk of *GUMOs*, or genetically *un*modified organisms, for the presumption will reign that organisms will be, normally, genotechnological versions of previously wild or domesticated species, and GMOs will be the standard not the exception. Current trends would seem to indicate that the ratio of GUMOs to GMOs, what one might call the *coefficient of trans-geneticization* in agriculture and in livestock breeding, will go on increasing, owing not only to the expansion of the front of genetic intervention, but also to contamination effects of wild or genetically unmodified populations by ever-spreading GMOs in farms, which would of course ruin the very possibility of continuing the experiments in biological farming, initially at least in some countries, where both are being carried out, and perhaps later throughout the world. Surprisingly, at the very time when *bioinvasion*, the harmful effects of exotic species wittingly or unwittingly transported by humans, is now regarded by many biologists as one of the greatest contemporary dangers to biodiversity, and public concern has been aroused on this score as never before, the new kind of bioinvasion, of insertion of exotic varieties – this time not of varieties brought in from elsewhere in the world's biota, but from the labs, the *genotechnological bioinvasion* of GMOs – is being consciously and strenuously propagated, and the fears of contamination of wild or domesticated species discounted with a brazenness that takes one's breath away. The prospect of the eventual disappearance of GUMOs, the advent of GMO world monoculture, with every species of complex organism bearing the hereditary impress of biotechnology, may be still on the distant horizon. Yet even if thought of as in the near future, it would not be one that, as far as can be seen, many in the dominant techno-economic elites would find at all disturbing or even conducive to a mild degree of uneasiness. On the contrary, many amongst them would find it quite pleasing, if indeed it is not seen as one to be brought about with all speed while the funds flow. As to the adverse, and possibly catastrophic, impact on biological farming, their silence is deafening. All one hears about from geneticists in their public communications is talk about the advantages of transgenics vis-à-vis the prevailing agricultural practices of extensive reliance on heavy use of pesticides and other synthetic biocides to secure high and profitable yields, that is, the comparison is exclusively made between two technoscientific or technological modes of agriculture within the normal economic context of corporate dominance, not between them and organic or biological farming. The latter of course is not at all anti-scientific or anti-technological per se (if the target of much hostility by many scientists, and at best regarded with suspicion, it has also attracted the support of some scientists for decades), or anti-capitalist though hostile to some corporate practices, but is certainly guided by a quite different spirit and according to different priorities. It has received little official support, unlike orthodox synthetic pesticide-intensive, biocide-intensive practices that have enjoyed a wide range of fiscal and other benefits for decades ever since it began in the late nineteenth century and especially after 1945. It is true that in the United States organic farming appears at last to be enjoying something of a boom, with steadily rising numbers of consumers turning to its products, which are becoming competitive even in terms of prices. There has been widespread revulsion against the massive use of synthetic pesticides in agriculture in the United States and their vast and persistent toxic legacy, and public opinion has induced some legislative and regulatory change, however limited and irregular, as the critics point out. But in the main, while the state may be neutral in many spheres, it has not been entirely neutral in this one. In addition to the support for orthodox agriculture, we have now the push for trans-geneticization of agriculture. This is aided by state facilities and coercion, not simply propelled by market forces, in a quite relentless fashion, clearly aimed at bringing about a large-scale, irreversible state of affairs in agrotechnology and agroeconomy. In addition to the genetic capitalism, which we discussed in connexion with the technology of human genetic modification, we are witnessing perhaps the greatest state-impelled reshaping of agrarian capitalism since the enclosure movement in eighteenth-century Britain – the emergence of state-backed, public opinion rejecting, technoeconomy promoting, *transgenic State agrocapitalism*, the maximization of transgeneticization, and the ubiquity of GMOs.

What we have called the 'ratchet effect' of naturalizations or of acceptance/validation for every consecutive emerging technology, especially RTs – which HRC defenders prominently appeal to as a knock-down argument while at the same time vociferously denying any commitment to technological determinism, and proclaiming its invalidity – would only ensure an unlearning, unself-critical, self-satisfied conservatism sacrificing the openness of the future (so prized by liberals, and quite rightly), to the alleged necessity of an unexamined, taken-for-granted even triumphalistically proclaimed recent past.

Eugenics

The eugenic movement, which flourished (under that rubric) throughout the West between the 1880s and the 1940s, advocated and worked strenuously towards major shifts in reproductive practices and institutions. At the beginning particularly, an important analogy in their elucidations was with the deliberate practices of plant and especially animal breeders; human society too should be a scientific breeding zone, and thus reproduction should not be left to chance, as it is not with racehorses and prize cows. It was driven by a sense of urgency in the belief that genetic 'stocks', ethnic, national, class-wise, perhaps the human biospecies itself as a whole, were deteriorating steadily under the circumstances of contemporary civilization, with the decline of the pressures of natural selection – though to what extent natural selection has been suspended in relation to our species is still a contentious issue. This *dysgenic master trend hypothesis* that our societies are not only genetically sick, but genetically deteriorating ('Degeneration') was a motif that was played in this period in every register, literal and metaphorical, physical, hereditarian, mental, intellectual, characterological, ethnic, cultural, ethical, civilizational, in every Western country; it was certainly widely propagated and stimulated in a variety of repressive policies. It was combined in the 1930s with the thesis of the alleged decline of measured national intelligence in the UK, or what one might call a trend towards increasing cretinization – both resembling a Marx-like directional law or trend of 'immiseration' under capitalism, in this case immiseration of the intellect.[30] This dysgenic master trend hypothesis continues to be reiterated as, for instance, quite recently by one of the most brilliant mathematical-evolutionary theorists of our time, and a virtually canonized ultra-Darwinian, W. D. Hamilton.[31]

This classical, or Mark I type of eugenics, envisaged essentially, *faute de mieux*,[32] changes in reproductive institutions, the laws, regulations and customs concerning who should be allowed to procreate, who should be persuaded to marry whom (on the assumption that marriage would be the reproductive site par excellence), who should be encouraged to have progeny or be persuaded not to, what types of procreative unions are to be favoured, and eventually in the late 1920s and 1930s the recommendation of 'children of choice' to be born from artificial insemination by superior donors to willing married women with conjugal consent, to add to the biological children of the couple, and so forth. They diverged as to the relative weight of the social mechanisms, state legislation and coercion, and suasive agencies (education, propaganda) that should be deployed in relation to what social strata. It should not be overlooked that even in democratic countries such as the Scandinavian nations, Finland, as well as some states of the United States, prisoners, and those confined to asylums, were favourite target populations for compulsory sterilization and even castration, physical or chemical, if not 'euthanasia', passive[33] or active. The term is often a euphemism, one of the most successful euphemisms of recent times, for the practices so described may well be involuntary, unsolicited, and not necessarily painless, and they might as well be called in many cases at least, *disthanasia*, as measures of coercive negative eugenics, practices that in some of these countries went on, if on a lesser scale than before, until the 1970s. They also diverged on the relative emphasis to be given on 'negative eugenics', the remediation and prevention of dysgeny, and on 'positive eugenics' to foster genetic improvement in successive generations, producing ever-higher average human types (in physical and intellectual, if not in moral terms) in the happily eugenicized populations. In the more fantastic versions, they traced their lineage to Book V of Plato's *Republic* and thus to some extent to what Dumézil called Indo-European ideology via Sparta, the most eugenic and female-equality attaining of the Greek polities, maintaining or upgrading superior castes or ruling races. Basically, the general tendency was to envisage negative eugenics as coercive, relying on the powers of the state to enforce sterilization of various biomedical categories such as the 'mentally deficient'. A bill was even put to the British Parliament to this effect, but was soundly defeated, though with the aid of fiscal incentives, cash prizes, honours, and so forth.

Although many geneticists (and other biologists) shared eugenic views, other scientists outside the life-sciences did also, as well as very significant numbers of physicians and psychiatrists entitled to, or charged with, signing and enforcing sterilization orders and death warrants for the elimination of the incurable, the severely disabled or mentally incapable (when they were not inducing abortions against the will of pregnant women classified as the bearers of dysgenic offspring, or arranging for their murders). This pervasion of eugenic views in the medical world partly explains the willing, even eager, participation of the medical profession in what happened in Nazi Germany from 1933 to 1945, and in Austria from the *Anschluss* to 1945, perhaps the single most-shameful period in the history of medicine, though it should also be recalled that the medical profession in Germany was already disproportionately Nazi-leaning even before 1933, to judge from their propensity, greater than any other profession, to join the Nazi Party before the Nazi seizure of power.[34] This eugenic climate of opinion affected the medical profession in all its branches in practically every European country, as well as the United States and Brazil. A eugenic outlook, or at least eugenic notions, affected the climate of opinion quite pervasively in this first eugenic epoch. Leading writers and intellectuals were sympathetic to eugenic policies and shared the eugenic vision of society. At some point in their careers social scientists – including economists of the stature of Keynes, sociologists (even left sociologists) and criminologists, a number of academic philosophers, some Protestant theologians, and a number of politicians – paid lip-service at least to the eugenic vision.[35] Immigration policies were perhaps the most affected (eugenic racism was an important syndrome of the time), and in domestic policies and practices in prisons, hospitals and asylums were also influenced by negative-eugenic recommendations. In the later 1930s, with the great spread of socialist and Marxist ideas in Europe and America in the wake of the Great Depression, a kind of non-racist, non-masculinist, Left or Social Eugenics, one might even say a Marxist Eugenics, developed with greater distinctiveness than ever before – recognizing that under conditions of great social inequality, and the social advantages of birth to the upper strata, which masked true innate talent such as prevailed under all social orders including capitalism, it would be very difficult to pursue eugenic policies fairly and effectively. But then the Marxist geneticist H. J. Muller was ready to instigate a large-scale programme of positive eugenics in the USSR, on the assumption that in the new supposedly socialist society, social conditions were now ripe for the implementation of positive eugenics, as true talent could now manifest itself. He seemed unaware that the USSR had been engaged for some time on a large-scale venture of 'negative selection' deporting, imprisoning for long periods or for life, or killing, large numbers of talented people as 'class-enemies'. In fact, this massive negative selection by official campaigns of genocide, class-persecution, ideological liquidation and the like, was one of the major features of the twentieth century.

It should be apparent in any case that eugenics was above all a *pensée agissante*, a movement strongly geared to action, policy, the transformation of societies and, in part, a reaction to supposedly dysgenic trends that could lower national intelligence and societal performance in international economic competition and war. Only after the discrediting of eugenic racism in the wake of World War II did some geneticists (as distinct from eugenicists without the same scientific qualifications) turn to what one might call retrospective, explanatory rather than policy eugenics, to the working-out in a much more concentrated fashion than ever before, of an eugenic or genetic theory of history as part of a wide-ranging biological history of the species in general and human populations in particular, but with special emphasis on genetic factors, especially with the eliciting of the latent genetic causes, the dysgenic explanations for the decay of cultures, nations, civilizations and the breakdown of empires.[36] The lesson they wanted readers to draw was primarily that the West was, by virtue of dysgenic trends, on course for decay and collapse, on grounds to be added to those of Spengler's racist one master argument that there was a virtual removal of the workings of natural selection in Western populations by virtue of the social protection of the inept and unfit, and the prevalent freedom of marriage and the freedom of procreation. This kind of topic has more recently, after an interval of some twenty years or so, become an area of inquiry filled by socio-biology and lately 'evolutionary psychology' with considerable fanfare, encompassing the entire history of hominids.

Like the eugenicists of the classical period between the 1870s and the 1940s, population geneticists continue to believe in the existence of a dysgenic trend in Western societies, the accumulation of deleterious mutations. However, the major difference between their current stance and that which prevailed in the classical period, is that they now claim, contrary to the urgent minatory tone of classical eugenicists, that this operates very slowly indeed, in terms of centuries and does not call for immediate massive remedial action. They conclude that, in the absence of sufficient knowledge about safe and consensual biotechnological intervention, it is best to maintain, by and large, the current genetic status quo until the distant future, when our genetic and psychological knowledge will have advanced substantially more: a 'hands off' stance for the time being, wholly at variance with the 'action this day' stance of their predecessors. This was the cautious conservative point of view expressed until twenty years ago, but now things have moved on beyond this cautious position, as we shall see, at least as far as positive eugenics is concerned. Negative eugenics, now largely absent from biomedical dealings with children or adults, operates (though not under this rubric) in our hospitals, with the practice of fetal euthanasia, in the light of prenatal screening, an area of very considerable technological advances in the last few decades, and presumably with the informed consent of parents, and even, on a much lesser scale, with neonate euthanasia.

However, the first or, if not the first then at least one of the first texts by a serious biologist to envisage not merely changes in reproductive *institutions* (marriage, family, family size norms) to accomplish eugenic goals, but also changes in reproductive *technologies* on a scientific basis, drawing on the findings of genetics, embryology and biochemistry, for the betterment of the species, was the geneticist, mathematical evolutionary theorist, polymath J. B. S. Haldane's challenging, even scandalous pamphlet 'Daedalus or the Science of the Future', published in 1924.[37] This was at a time when dozens of little books were being published on the future of this and the future of that, but this one was certainly the most remarkable, and the most remembered.[38] In it Haldane proclaimed the coming of the age of 'biological inventions', after the great physical and chemical inventions that had been changing the world of manufacture, agriculture and communications, a new phase of technology or the scientific transformation of things and persons, with tremendous social implications, which age has of course come to bloom in our days, although no doubt in many ways there is much, much more to come. The constitution of eugenics had preceded the reception if not the formulation of Mendelian genetics by a few decades. Galton's founding text dates from 1867, and the Mendelian regularities in the appearance of characters across generations, though formulated in the 1860s, were only accepted by the scientific community much later, after 1900.[39] Some eugenicists strongly opposed Mendelian genetics well after it had become widely accepted amongst geneticists and other biologists; a case in point was the leading statistician and biometrician Karl Pearson.[40] Nevertheless Haldane not only accepted Mendel–Morgan genetics, but tried to outline how it and other recent developments in biology could serve to develop new RTs; thus eugenics need not rely solely on directed social change and public policy to accomplish its aims of correction and improvement to the genetic basis of human populations, to steer national biological prospects and indeed human macro- or mega-evolution on the planet.

Haldane's pamphlet is no mere historical curiosity, for it had an immediate impact on the chattering classes. It was reprinted many times in the following years. Among notable responses that ensued over the next five years or so in the UK, it drew the interest of one of the first women journalists in Britain to write on science questions and to go on to write a science-fiction novel inspired by the problematic pamphlet (recently reprinted and discussed by feminist scholars),[41] and to marry the author (she became Charlotte Haldane). It triggered a response in the form of a pamphlet entitled 'Icarus or the Future of Science' (the very title indicates that it was meant as a counterblast to J. B. S. Haldane's work), which later developed into a full-fledged book by no less a figure than Bertrand Russell, and was one of the major influences on Aldous Huxley (brother and grandson of biologists), whose novel *Brave New World,* was published in 1932.[42] It addressed the role of RTs, especially ectogenesis (the entire reproductive process from fertilization to birth being carried out in vitro, outside any human or animal body), which Haldane had envisaged, and the factory production of fungible, interchangeable, genetically identical human beings. Though they were not clones in the precise contemporary technical sense in that they were not produced by somatic cell nuclear transfer, they were co-natal, unlike cloned and clonees today, or prospectively, who belong to different generations, by a kind of *genetic Fordism* as well as psychological, psychotropic and social ones in a scientifically planned society.[43]

Haldane's pamphlet was significant in stressing, perhaps for the first time, the new role of biological inventions in shaping human society, the role of what we would call today biotechnological innovations in animal husbandry, agriculture, public health and nutrition. Conventional thinking about the future of the economy was still geared to physical and chemical inventions in both peace and war. Indeed, World War I has often been characterized as a chemists' war, and World War II as a physicists' war, though biological weapons were used by Japan in China, and chemical weapons descended from World War I in the subsequent Holocaust, but also more directly and intimately, for the basic structures of human reproduction. Though he missed the mark in expecting ectogenesis to come any time soon (it is still not contemplated for the foreseeable future, and it may be still quite a few decades ahead, though interest in its feasibility has re-emerged) and represent the major RT in the biologically informed society, one could argue that HRC is the 'next best thing' to ectogenesis, inasmuch as it constitutes a radical departure in the basic form of human reproduction from the inception of the species, if not since the inception of sexual beings in the history of life – accomplished first, in the case of mammals, with Dolly, at least with a long-lived specimen.[44] It should be noted that he saw the coming age of biological inventions not merely as biologically beneficial for individuals, or for the species, but also as happily scandalous, pre-eminently morally and intellectually purgative, particularly in its 'demonstration effects' undermining religion, the authority of priests and of received moral and legal codes regarding sexuality (from which codes he suffered personally and professionally), liberating the whole realm of human sexuality and reproduction from the extant of conventional non-technological constraints (Haldane 1924). It would, as it were, emancipate the forces or means (the new technologies) of reproduction from the extant social relations (i.e., the moral, legal or religious frameworks) of reproduction, separating not only sex from reproduction but also reproduction from sexual activity, if not in the then unforeseen HRC, at least possibly in parthenogenesis or ectogenesis, and both sexual reproduction and sexual activity in general from legal compulsions, at least beyond some absolute minimum.[45] For our biologist these biological inventions would act, as was said of the telescope in the eighteenth century, as 'moral artillery' to blow up the spiritual heavens.[46] Curiously, the microscope was not generally assigned the same role, though Haldane went on to live and work in India partly because it was a neutral country in the Cold War, expressing sympathy with the Hindu way of life.

Like many other technological innovations HRC will not be brought about by market-pull, let alone by the press or call of human need. If anything, it will be a matter of technology-push, or the drive to realize what may be technologically possible as such, and the credit, fame or notoriety accruing thereby to the first in the race, while justifications of other sorts can always be found aplenty after the event, even perhaps a market rationale, legal endorsement or non-prosecution as we have seen in previous new RTs. Thus, what might be called '*the propaganda of the technological deed*' has been more successful, more widespread, more condoned and more contagious than the tactics of the political, often violent, 'propaganda of the deed' adopted by the anarchists in the late nineteenth century, which was generally counter-productive, most of all in the case of RTs. Any moral or legal barrier to new RTs can be breached somewhere, involving at least 'symbolic violence', profanations, for some religious perceptions of the world and, once breached anywhere, it is irresistible everywhere else, with perhaps only a short lag. It remains to be seen whether this general effect will or will not apply in the case of HRC. In the case of the cloning of pets – 'members of the family' so to speak – some effective demand has already been expressed, and down payments have already been made to at least one company in the United States, but so far no deliveries have been reported; it appears to be a perverse conspiracy of circumstance that the bulk of the animal species indulged as pets by humans today, at least in the West, are not amongst the most easily cloneable species.[47] Nevertheless, the prospects of HRC must be seen within the structural–historical context of 'techno-capitalism' in which technological change has become recognized as the motor par excellence of economic growth, surpassing the other factors that have hitherto concurred with equal or comparable weight. Soon, this trend might incorporate, perhaps, as an important sub-sector, a *genotechnocapitalism* as marketed, implemented techno-biological innovation becomes a salient and lucrative area of academic–industrial science capitalism or 'the techno-science of commodities'. Since the production of commodities is accompanied by the production of 'discommodities', it is also the 'technoscience of discommodities'. This surge involves the current explosion of genetic and genomic information and information-processing capabilities induced by the race of the Human Genome Project, accelerated even more from outside the original research community by the likes of Celeron. The commodification of 'epistemic things' in genetics and biotechnology, the thrust of genetic engineering in all its phases includes not only somatic (somatic gene therapy has turned out a great disappointment, though this may be changing), but also germline genomic engineering. The emerging global genetic market is accompanied by resistance there in the Third and Fourth Worlds, and elsewhere to the idea of a potential global human genetic market, a resistance that seems to have taken the geneticists involved completely by surprise.

A Typology of Eugenics: Mark I, Mark II, Mark III Eugenics

Human Being as His/Her Own GMO: The Coming of (Trans)Genetic Capitalism

What has come into being is what may be called Mark II eugenics.[48] This may be characterized in terms of the following criteria, which differentiate it from classical or Mark I eugenics as it flourished between the 1880s and the 1940s (sometimes equated with eugenics *tout court*):[49]

- Mobilizing the more recently developed technologies of genetic engineering not available in the epoch of Mark I eugenics, rather than aiming at changing social institutions and legal regulations in order to bring about the right quantity and quality of procreation (by eugenic and demographic criteria) directly, as its prime instrumentality.
- It does not appeal formally and directly to the state, though presupposing state consent for genetic modification of persons ex ante or ex post, in a 'culture of rights',[50] which will now include genetic rights – rights discourse being, ostensibly, virtually the supreme moral idiom of the epoch, whose inflationary tendency has not yet come to an end. Of course, rights will have to be enforced by the state against moral minorities, or perhaps even majorities. It should be noted though that in practice, one or another version of utilitarianism, as much as, if not more than, deontic ethics, shapes or justifies the big decisions involving economy and technology through cost–benefit analysis and evaluative procedures of varied kinds.

By contrast, Mark I eugenics stressed above all positive and negative duties and obligations, the reproductive duties of parents to their children (including the duty not to procreate in, for instance, cases of hereditary illness) and of prospective parents to their country and its future in a competitive world, and of all (including the state) to future generations, duties and obligations which would have been normally construed as of supererogation, but which the Mark I eugenicists would want to make strict duties and as far as possible legally enforceable as well as backed by enlightened public opinion. It cannot be said that they were reluctant in this, for they were enthusiastic propagandists, even resorting to the medium of films to get their message across, though they were also keen if not keener to accomplish (macro) eugenic tasks from above:

- Not targeting, in general (or not formally), special social strata, or other collectives and thus deserving to be called *micro-eugenics*,[51] or individualist eugenics, legitimated by the language of individual rights and of personal choice, in a climate of increasing geneticization of our self-understanding and self-explanation.[52] Supposedly, we are what we are largely by virtue of our genes. 'Genetic determinism' may be true or false, and this leaves open the question of what sort of determinism may be implied, classical or 'chaotic'. But the mode of monocausal attribution of specific human traits to specific genes, in a simplistic one-to-one fashion, is widespread, reinforced by monthly – indeed weekly – announcements in the media by innumerable scientific teams of the identification of the gene for this and the gene for that behavioural characteristic or faculty of mind or even disposition of belief. The consequence is

> that the propositional schema 'the gene for [. . .] such and such' is constantly refilled with new matter, bracketing the role of other factors in gene expression and epigenesis, which would have qualified the geneticism. It should be added that the diffusion of neuroscientific or neurological determinism is coming in fast, not least with the publicity coup of the promulgation of 'neuro-theology', with astonishing public announcements of the neuronal arrangements underlying religious belief or mystical visions, reminiscent of nineteenth-century phrenology. Instead of a gene for God, or a God-bump as a phrenologist might have posited, neurons for God! 'Our fate is not in the stars, but in our genes', uttered James Watson, of double helix fame, great instigator of the Human Genome Project and, indeed, to a degree of the financing boom of bioethics. Thus, the genotypes would be the counterpart to the astronomical conjunctions of genethliac astrology, to pursue the analogy, and genetics a sound counterpart of natural magic.[53]

We could not supposedly change our star-imposed birth-fate in the framework of the long tradition of thought in various civilizations, and no less influential in the West, which has been called 'astro-biology'.[54] Historically, human beings are often most activist, world-conquering and world-transforming when they have professed forms of theological or secular–historical determinism, as in early Islam, Calvinism or Marxism. Now, with contemporary genetic technology, instead of resigning ourselves to a kind of 'astro-genetics', in which the genes play the determinative, life-shaping role once attributed to the stars, we can – or will soon be able to – change, our genes, or a subset of our genotypes. Thus, professed genetic determinism is being rapidly complemented by the practice of 'genetic interventionism', at least for a subset of genes, and thus the determinism in question, for this subset (and this of course an important qualification), turns out to be a 'soft determinism'. This exemplifies Collingwood's philosophical thesis that, in practical inquiries, what is identified as a cause, is precisely, and solely, that which we can successfully manipulate in order to bring about a desired effect, a thesis sometimes called the 'handle' theory of causality (1940).[55] This (partial) genetic interventionism would be achieved particularly through the introduction of super-alleles – at a price, with the right technology, with more investment in science. This is always the subtext, when not explicit. We would thereby change our birth-fate and presumably wholly for the better, though perhaps the word 'destiny' would have been more fitting than the word 'fate' in this context. Or rather, we will find ourselves, in due course, re-fated, through the social implications of genetic intervention, condemned no longer so completely to the 'lottery of birth' ('the stars', 'the genes'), but condemned instead to the *treadmill of permanent genetic improvement*, for the lure of such genetic improvement will prove irresistible to some, and eventually to all, as it is quite likely that virtually no one, whatever their religious or moral objections or misgivings, will wish to leave their children, or even themselves, behind, in the race for genetic enhancement of body and intellect. Envy, that terrible demon that has haunted utopianism, would under such circumstances, take the form of 'super-allele envy'.[56] Genocracy does not require direct rule by geneticists, genetic engineers of the cruciality of their enterprises, but the maximization of the scope for genetic/genomic intervention, and the implementation of such potentialities – and for this purpose it would suffice to instigate and set in motion the treadmill of genetic improvement we have just indicated, a new and more profound mode of the over-all, global 'technological treadmill' (as various authors have called it) on which we are condemned to live until the end of the species.

- Market orientation: in contrast with Mark I eugenics the state is not, and should not be, a player. Eugenic action serves the new market for high-tech genetic or reproductive services that have emerged and developed in increasingly sophisticated ways in the last thirty years or so, in conformity with the market ideology of the Republic of Choice, in which the presumption is that nothing is given and fixed, everything is alterable by technoscientific intervention and everything that matters must be chosen freely by individuals with a minimum of state constraint, which is attaining, given the conjunction of the ideology and the technologies, perhaps its last and highest stage as a Republic of Genetic and Transgenetic Choice.
- The macro outcomes in the reshaping of at least national society, as a result of multifarious micro-eugenic, genetic interventions in individuals and families, or the possible negative sociocultural 'externalities' of such geno-economic activities, are deemed to be beyond its purview, or indeed anybody's purview in a free society, while they were the clear public target, and indeed the chief aim, of Mark I macro-eugenics. 'Reprogenetics', or germline genomic engineering, encompasses what used to be called negative and positive eugenics, including therefore enhancement, perhaps even the indefinite enhancement of every quality and attribute, physical or mental, of human beings with some kind of genetic basis.[57] Such an improvement can be compounded or reinforced even within a lifespan by activation or replacement of the 'superalleles' installed in the prenatally genetically improved subjects, so as to upgrade her or him genetically within his or her lifetime over and over again. Insofar as genetic improvement can now in principle take place in the individual's lifetime in those who request and pay for it, we have now reached the stage in which *self-eugenics*, the genetic improvement of one's 'natural capital' or 'natural' inherited assets, the project of, so to speak, designed natural capital accumulation of the genetic self, of revising one's genetic biography, or extending one's genetic curriculum vitae, may become possible. Sartrean existentialism never imagined this type of self-reinvention, which must surely count as inauthentic by its criteria.[58] Here, the concern is strictly with the self – not, as in classical eugenics, with the common good particularly of nations, or with public policies involving great numbers, or taking into account the national population as a genetic collective. The emergence of self-eugenics, of each human as his or her own ever-targeted GMO, his or her own laboratory of genetic self-modification, would make for a further differentia for Mark II eugenics, or shall we call it *Mark III eugenics*?

All this, of course, is in addition to the other techniques of *euphenics*, not only the correction of disfigurements and other remedial measures, but also the improvement of one's phenotype, especially the enhancement of one's looks, through such means as cosmetic or plastic surgery, which has been typically individualistic from the beginning, neither aiming at social change nor part of social programmes. The British novelist L. P. Hartley (better known for his other works, some of which have been made into films), published what one may call, despite the oxymoronic ring of the phrase, an *euphenic dystopia,* a rare instance of this little-frequented minor genre, entitled *Facial Justice* (1960). Reacting against the egalitarian ethos of British society in the Labour Government years of 1945–1951, he pictured a society in which everyone's looks are standardized by compulsory surgical interventions, so that no one need suffer from being ugly, and no one need be envied or admired for being beautiful, the ideology being that one's good looks are totally unmerited, and yet the distribution of looks is very inegalitarian and therefore calls for correction, just as other unjustifiable social and economic inequalities have to be corrected, as a matter of social justice, through fiscal and other public policies (Hartley 1960). The novelist probably imagined that this kind of intervention would never become widely available or widely desired, and he essentially envisaged socialist ideology as the sole promoter and the state as the demiurge, and the only demiurge, of this kind of transformation.

Nevertheless, waves of euphenic fashion and the pressures on or demand by females and males (on balance probably more by females) in some age cohorts, at least in some countries, for such interventions, given their increased availability, safety and sophistication and the extension of the legal market for euphenic services, may bring about not insignificant social outcomes. Some have already expressed concern that the potential Republic of Euphenic Choice may bring about not so much a freedom or chaotic distribution of looks, as a kind of 'MacDonaldization of looks', as it was phrased recently in a debate promoted by the periodical *New Scientist* in the UK. Technological optimists, as is their wont, might well characterize such a situation in a quite different way: say, as one of the elimination of deformities, of ugliness, the opportunity for the exercise of self-choice, of self-invention, a kind of biologization of Sartrean existentialism, and the rise and rise of facial, somatic, euphenic standards, although aesthetic relativists or aesthetic dissenters might also complain about the uniformity![59]

Like every other technology that has come into play in human societies, including biotechnologies, HRC is likely to interact with other technological or techno-economic developments. We have become familiar with the press of marketization and commodification of human-biological constituents, of organs and body parts (what has been called 'the universal body shop'), assisted reproduction services and genes. In an analogous fashion, '*the universal gene shop*' or the 'world gene market' to come, for such 'gene shops', 'gene markets', or the 'genetic supermarkets' to which the political philosopher Nozick referred as a Very Good Thing, still draw from restricted parts of the world. And one has to mention also the generalized disposability of frozen human embryos, sperm and oocytes, for a variety of purposes, including the increasing emphasis on what has already been called 'techno-sperm', and surely the 'techno-ovum', cannot be far away. The 'techno-womb' has already been envisaged, and more recently the cloning of embryos for therapeutic purposes, which is now a major theme and even demand of many research workers as absolutely necessary for medical research and therapy, disregarding the claims by other scientists that adult cells may equally well serve the purpose and may therefore obviate the need for such human therapeutic cloning. In the moral climate associated with and fostered by these trends, it is problematical what strong barriers the universal technicization and commodification of human-biological life may encounter.

To appraise the project of HRC solely in terms of the 'preferences' of the to-be-cloned seems quite unrealistic, quite apart from the fact that, despite numerous disclaimers, there obtains a strong proclivity in (micro) economorphic modes of reasoning, to imply that they may be taken to be determinate, given, fixed, imperious, self-contained, consistent or transitive, connected and complete, an ensemble of properties that might be called the ideal type of *Platonic Preferentialism*, favoured at one time by neoclassical economists of the Chicago variety. Note though that this dogma was always rejected by the Austrian variety of neo-liberal economics which leans towards a *Heraclitean Preferentialism*, with preferences always changing from situation to situation, never complete or necessarily transitive, or well-defined, though both subscribe to the dogma of the Absolute Individual and the Sovereign Consumer.[60] Now the Sovereign Consumer feels entitled to access the market for genes, as any other market, the market for genetic therapy, the market for enhanced 'genetic capital' for self, and especially for correcting and enhancing the genetic resource endowment for progeny, this presumably being repeated in generation after generation, a process of genetic capital accumulation without definite predictable limits in which Genetic and Genomic Capitalism (if you like, G2 Capitalism) may advance; proteomics may join this process in due course.

The project of HRC appears to embody the conjunction (in some ways rather uneasy), of two great dreams, one very, very old, the other having more recently come to the fore: the dream of the Perfect Copy and the dream of the Perfect Baby, combined in the Perfect Being (Agar 2002). To be sure, in an age of mutabilism, or pan-fluxionism, of gales of creative destruction, HRC is likely to yield to the more compelling vision of the *ever more perfectible being* via biotechnologies of various sorts (genetic or genomic, both germinal and somatic, neurological), though it may coexist with it in the pluralistic Republic of Genetic Choice that is emerging. Of course, it can harness other motivations, draw on other dreams, and in accordance with the principle of the 'heterogony of ends', it is likely to. It is one, but only one, of the many ways in which what we may call *genic* (or *trans-genic*) *perfectionism*, perhaps the single most salient, most appealing, most credible mode of perfectionism in the West today, after the discrediting of all other radical social and political versions, heir to all the ruined visions of radical and enduring, life-long and trans-generational, improvement ('enhancement') of human beings, is evolving.[61] Perhaps we have not really learnt to live without illusions of omnipotence and intra-mundane redemption, as all the sages have taught us to strive for, if we cleave to genic or trans-genic perfectionism, or any other kind of perfectionism, as our royal road out of our miseries. It is a perfectionism geared to individuals and individual choices, suited to an age of atomistic individualism. Prima facie, it hardly seems consonant with the interests or affinities of *homo heterarchicus,* though both partake of digital information technologies as instruments. Technologically induced self-transformation, second-order or reflexive anthropogenesis writ small and piecemeal in the manner of genomic engineering, seems the way ahead under these circumstances.

Folk reactions to the prospects of HRC have been of great concern, even repugnance, mixed perhaps with a sense of resignation. But then, 'folk bioethics', if we may speak thus by analogy with the consecrated locution 'folk psychology', does not enjoy a good reputation, just as Descartes and Malebranche despised the common-sense prejudice that animals – any animals – could suffer pain, embracing the rational–scientific view that they were but natural automata. The concern and the repugnance appear to be eminently sound, and I entirely concur with the extraordinarily eloquent and insightful paper by Leon Kass, entitled 'The Wisdom of Repugnance' (1997), on this and other matters.[62] What about the resignation that is wholly understandable in the light of the flood of current geno-hype and RT hype, and the ever-greater geno-techno-market incorporation of all branches of life (human and nonhuman) is the subjugation of more and more areas of privacy, intimacy and personhood to 'technologico-Benthamite' canons.

In metaphysical terms, a Principle of Plenitude has operated in Western thought, at first as a celebration of a fixed, value-laden, teleological Order of Nature, then as a heuristic for an all-encompassing evolutionary process and as a 'teleological principle' of the sciences in general, or as a 'principle of reason' more recently as a charter for technological development, though still important in cosmology and modal thought, as in the analytical metaphysic of modal realism advanced by David K. Lewis (1986). In the latter form it is also known as the 'technological imperative', or in more restricted and mundane fashion as the 'Gabor rule', as well in other guises such as 'can implies ought' (a reversal of the familiar Kantian ethical principle), identified by the systems theorist H. Ozbekahn in current technology worship. We can either let the Principle of Technological Plenitude work its way through or we can attempt to draw boundaries; there is an element of arbitrariness in any drawing of boundaries, but this does not make them necessarily indefensible or illegitimate, for any liveable life has to be sustained by boundaries not given obviously by nature, though this liminality of the passage to asexual reproduction for human beings and the deliberate production of genetic copies, even if perfect seems more luminous than most in the matters of RTs, a kind of *tremendum*,[63] a kind of utterly misplaced venture into the 'technological sublime'. Of course, the claim that it involves a liminality, the crossing of major metaphysically significant threshold, a shift in some basic parameter of human existence would be totally rejected by many if by no means all geneticists and their fellow-travelling bioethicists, though it is not at all clear what, if anything, done or projected by scientists, could possibly count as such for them – such a category seems an 'empty box' for them. It would be foolish, however, to see this value and hermeneutic gap, as the clash of two camps, for science and against science, the enlightened and the obscurantist, the rational and the irrational (fearful, anxious, gloomy, much too ready to predict perverse outcomes and dangerous political utilizations of sound technologies). Rather, to paraphrase Whitehead, the mark of a civilized person is to stand unflinchingly for and by some distinctions and boundaries, on issues that he or she cares deeply about, however precarious and challengeable, even if he or she expects to be defeated, if only to bear witness.

In the same spirit, we may cite the religious ethicist Paul Ramsey, who already decades ago engaged presciently with some of these questions concerning the ethics of RTs. One of his books was already entitled *Fabricated Man*, in which he observed that 'a man of serious conscience means to say in raising urgent ethical questions that there may be some things that men should never do. The good things that men do can be made complete only by the things they refuse to do' (Ramsey 1970, 123). Or, in other words, human rational autonomy, and the human capacity for good, call as much for *nolonté* (Renouvier's serviceable term), as for positive volitions as *volonté* (Renouvier 1854). This runs so much against the spirit of the times that it is unlikely to be heard, let alone heeded. Never more so than today, at the juncture of liminalities in reflexive second-order anthropogenesis, the technological modification of the human being by (some) human beings, who have assumed, so to speak, not without arrogance, even with hubris, what we may call *the genetic general will of the species*. It may not be too far out to call their projects a form of soft genetic anthropological Jacobinism, which may yet take over the Republic of Genetic Choice. 'Germinal choice' was apparently an expression invented by H. J. Muller some decades ago, when trying to set up his Californian sperm bank to impregnate willing women, married or otherwise, with the seed of geniuses, or at least Nobel Prize winners, after his great disappointment in the Soviet Union when the all-powerful Stalin did not take seriously his eugenic programme.[64] Jacobinism involved the fusion of the executive, legislative and judicial powers; it refused any limitations of power according to a constitution – indeed it refused any limits in the name of the people, as in this case genetic Jacobinism, *the genocratic drive*, the drive to transform humans, as well as other kinds of living beings, according to genetic/genomic science and engineering (which is the form that the eugenic drive for biological/reproductive reform of society as a whole has increasingly tended to take after the acceptance of Mendelian genetics and especially with the advent of molecular biology), refuses any limits, any constraints, in the name of science or progress, or the rights of the Absolute Individual, dogmas which, in effect, seem to replace that of the sovereignty of the people, indivisible and inalienable, the dogma to which so many constitutions still pay lip service.[65]

'Science speaks to power': some geneticists/eugenists of standing[66] wrote to Mussolini or to Stalin (as H. J. Muller had the temerity to do, while living in the USSR, though he never got a reply),[67] begging them to implement without delay a state programme of eugenic reform in their countries, and would undoubtedly have written to Hitler, too, but for the obvious fact that he did not need to be written to. In any case, his putatively eugenic deeds were much commended by at least some geneticist/eugenists of professional distinction outside Germany, as well as by non-scientists favourable to the eugenic project. One example was the famous playwright G. B. Shaw, one of the early members of the Fabian Society, in his letters of the time to his fellow-Fabians (indeed almost the quintessential Fabians), fellow-socialists and by then public admirers and advocates of the Soviet Union under Stalin, the Webbs. Shaw had long been interested in biological speculations and the theory of evolution, especially as applied to human beings, as is apparent in some of his plays and their long prefaces.[68] It is true that like some other Left Eugenicists he regarded the equality of income as a precondition for eugenic intervention, for otherwise it would be distorted by the existing class system, but that did not stop him from approving, at least informally, negative eugenics policies even in very unequal capitalist societies (as was the case in Nazi Germany).

Today, fortunately, all you need, in principle, is to work through the market (though there can be no doubt that if political power were to demonstrate its willingness to help things along, even with the use of coercion, this would be most welcome). For this all you require is the sacred name of Choice, once sufficing as freedom of occupational choice and freedom of consumer choice, in which goods and services available for consumption were very far from encompassing the new choices regarding self (germinal choice, genetic choice, genomic choice), and with the immense help of profit-seeking ventures and corporations, perhaps one day these will fuel a major sector of the New Economy. The wondrous, cornucopian, Republic of Choice becomes an even more perfect Republic of Choice, by becoming a Republic of Germinal Choice, a Republic of Genetic and Genomic Choice with genetic/genomic supermarkets and hypermarkets – increasingly approaching a world scale – added on to all the others. You do not call this social engineering (such an ugly locution!), though genetic engineering or genomic engineering are noble things, since they consist of scientific, technical and social progress accomplished and distributed via the market, without state direction, so that it is perfectly all right, and legal to boot. Indeed, if you can do it this way; all you need is the non-interference of the state or indeed anybody else (religious, moral, intellectual obscurantists, anti-science survivors, take note, and keep away!). Jacobinism via the market! The Jacobins missed their hour, but the biocrats or genocrats of the world have met their hour of destiny, not in a dictatorship or totalitarian system. Admittedly, they did try their best in such regimes to implement eugenic – especially negative-eugenic – programmes, and would very likely do so again, a few possibly with some pangs of conscience, but progress cannot wait when the hour is ripe, not via state legislation in democracies, but at least in the first instance, in market democracies which, in the present conjuncture, afford the greatest window of opportunity for biocratic, positive-eugenic and genocratic projects, for the diffusion of 'genetic/genomic capitalism'. They are unlikely to miss their chance this time, with greater knowledge and, especially, far better tools than they had in the 1930s even though eugenic states appeared on the horizon, or indeed since the first eugenic law was passed in the United States in 1907 (rapidly followed by a number of other countries). Eugenic legislation was of course very largely negative-eugenic and positive-eugenic measures, which were usually indirect. Now the eugenic market society, or perhaps one should say more exactly the *micro-eugenic market society*, with a heavy emphasis on positive eugenics, replaces from the point of view of the genocratic project the failed models of the *macro-eugenic state capitalism* (of which Nazi Germany was an extreme case), and the *macro-eugenic state socialism* that Muller (and other biologists inside and outside the Soviet Union) thought possible, desirable and urgently necessary (Muller 1935).[69] Their hour of destiny, the hour of controlled geneticization of humans and society, of maximum genocratic drive, is also our hour of being subjected to, or indeed of surrendering willingly to their designs, their siren songs.

We will be the agents and patients of yet another phase of the anthropogenic *Experimentum Mundi*, in which the entire Earth is subject not only to unwitting human impacts but to deliberate human intervention and to corrections of the results of previous human interventions, in a never-ending spiral. It seems the best that can be hoped for is to turn it wholesale into what one might euphemistically call an all-inclusive Garden, wholly in our care, reliant solely on our oversight, benevolence, knowledge and wisdom – not to mention cosmic luck (this has been called 'gardenification' in the literature). Something very like this was envisioned by Leibniz more grandiosely perhaps than even the other geo-theologians of the seventeenth century, some of whom regarded the Earth of their time as much-decayed, largely unproductive, and not least very ugly. Mountains, for instance, were seen as per se utterly repulsive, as telluric disfigurements, in sore need of thoroughgoing improvement and beautification everywhere. Geo-theologians envisaged the restoration of things before the Fall as the *telos* of human action, of scientific and technological enterprises, of human history, in the earthly theatre (Wakefield 2010). Leibniz was eventually followed in this vision by others but particularly perhaps by Spencer (1857). Both were inveterate cosmic perfectibilists and devotees of the beneficence of human progress, though the latter without the guarantee of an all-good providential Deity. There would be a new sense for the motto *cultiver notre jardin* of Voltaire's anti-Leibnizian satire,[70] for now it would be a necessity and a strictly collective obligation. *Notre jardin* will be the whole Earth, not our little individual refuges, apart from a few flowerpots. Once upon a time, they used to say in the eighteenth century (for instance Locke in his political philosophy), 'all the world was America', meaning all was wilderness. Now for the first time we can envisage, with some degree of realism, a situation in which all the world would be an anti-America, as all the world would be 'garden'. A Garden without a Wilderness as its counterpart or surround, as was not imagined even in the great twentieth-century technological dystopias. Given the pervasiveness of technological transformation, monitoring and control of all earthly things, it would no longer be appropriate to speak in this case of the feasibility of the ideal of the 'machine in the garden'.[71] And that garden in the nineteenth-century American literary imagination, presupposed a wilderness, or 'wild' regions of considerable size as a counterpart, of a technological pastoralism, a vision that far outweighed Jeffersonian agrarianism in the United States. All we could hope for would be the *garden-in-the-machine* or the *machine-garden*, a garden maintained, policed, made possible by and saturated with machines and technological systems on every scale, or an interstitial or residual garden in a machined world, and not all the latent biophilia imputed to innate human nature by some sociobiologists (E. O. Wilson) would suffice to prevent this outcome. Thereby indeed the locution 'Earth Machine' will acquire a new sense.

More specifically, in a supposedly post-utopian age we will be our own guinea-pigs, the guinea-pigs of the self-induced *Experimentum Humanum*, the reflexive anthropogenesis accomplished in part through genetic/genomic science and engineering and, perhaps in the near future, also through regenerative medicine and neuro-engineering, not to mention the ongoing work in AI and AL (Artificial Life). In such an 'experiment' in wholesale and irreversible anthropological engineering, the technoscientific *homo hominans* comes to prevail over the 'natural' *homo hominatus,* or at least will set the agenda. It seems to imply megalanthropia, the vision of a vastly superior human kind, via genetic/genomic modification, humans as their own GMOs, and soon only human GMOs or, for short, *GMHs*, genetically or 'reprogenetically' modified humans, with the *GUMHs*, the genetically unmodified, 'wild' humans, eventually a kind of marginalized and biopsychically underequipped Neanderthals of our time, fated to extinction (though some would perhaps be kept in an 'anthropological zoo'). No doubt in a world in which the class of GMOs would tend to exceed more and more, the class of GUMOs, the genetically unmodified organisms and especially the GUMAs, the genetically unmodified animals, would perhaps be confined to wildlife parks, and perhaps the current acronym GMO will eventually fall out of use, as obviously redundant, since the standing presumption will be that any organism, unless otherwise noted, will be a GMO, and GUMO will replace it to indicate those that, as far as is known, have not *yet* been genetically modified.

But more likely, as many others of varied ideological persuasions urge and prophesy, beyond the GMHs we will have 'mixed humans', 'chimerized humans', 'post-humans', 'meta-humans' or 'ultra-humans': cyborgs, cybots, bionic beings, chimerized human-animal beings (via xenotransplants), or chimerized human-animal-electronic beings, electromechanical-biological compounds, androids, robo sapiens, and nanobots. There will likely be a vast ultra-anthropological spectrum, inspired in part by a kind of *ontological misanthropy*, by which I mean a loathing of the human as such, in every way, in the grand scheme of things, without the theological safeguard of the aversion to generic human depravity that can be found in Calvin, for example, and therefore congruent with the will to the abolition of the human – a curious horror of the flesh, of 'fleshware', 'meatware' or 'wetware', as some of these authors (such as the prolific computer scientist M. Minsky) like to say, to convey their contempt for our anthropocentric or incarnate conceits; in fact, the two visions are not wholly disparate.

Human 'genetic capitalism', is the genetic, especially germinal, modification of human beings effected by the technologies available through market supplies and forces (the last and highest phase of capitalism? Genocracy: a superior, more inclusive form of market democracy? The Republic of Genetic Choice, the consummation of the Republic of Choice?). At any rate, these considerations furnish not wholly implausible scenarios in the light of current enterprises, prospects and speculations. Transgenic agrocapitalism, as we have called it, is certainly well on the way: state-backed, an alliance of political power, corporate interests and technoscientific knowledge not seen since the beginning of the Nuclear Age and its enormous unfulfilled promises of abundant, virtually free ('too cheap to meter'), totally safe energy for all, and for every purpose.

Notes

A *very* much shorter version appeared in English and in Portuguese under the title 'The Liminality of Human Reproductive Cloning, or In Praise of the Wisdom of Repugnance' in *Interact* (Martins 2002).

1 An example of such as the reported remark – perhaps not entirely facetious – is by the biologist and science writer Richard Dawkins – that he would like to have many clones of himself. Facetious or not, these remarks, and those of other scientists, as reported in the press, betray a disturbing vulgarity of mind, hardly conducive to a 'public understanding of science' or enhanced public respect for scientists. Another example of vulgarity on the part of scientists, this time by those actually involved in cloning, occurred in the naming of Dolly.

2 The use of the word *cloning* in this context has become generalized only in the last two decades, for the word *clone* came to the attention of English language dictionaries of new words in the last couple of decades first as an item in the vocabulary of computers; 'genetic copying' was used more often than cloning before in scientific popularization. A biologist reporting on accepted scientific usage distinguishes four senses of cloning in the literature: molecular cloning (making copies of genes), cell cloning (producing cell lines with identical properties), embryo

splitting (not true cloning) and nuclear replacement transfer, or cloning *stricto sensu,* involving cells or organisms, in the former case known better as *somatic cell nuclear transfer* or SCNT (Jones 2001, 25).

3 By 'conventional bioethics' I mean those recent writings in bioethics that have most impinged on me. I am sure I am not being fair to a lot of bioethicists, for I do not claim to have perused a representative sample of the corpus of bioethics, though I am also sure that they apply to some, and not the least vocal of the writings in question, and that is why the term 'conventional' may be applied here.

For the concept of 'evaluative metaphysics', coordinate with the more familiar pair of categories of 'descriptive' and 'revisionary' (or 'prescriptive') metaphysics formulated by Strawson; see Nicholas Rescher's *Essays in Philosophical Analysis* (1969). An almost identical conception of this third mode of metaphysical reflection had been advanced by another American philosopher, W. P. Montague (1925). It may be noted that another kind of metaphysical reflection besides these three was that advanced by Collingwood, for whom metaphysics was essentially a historical discipline, studying the 'constellations of absolute presuppositions' that have underlain the modes of thought of successive epochs (1946). He had in mind Western civilization. However, it is not clear that this kind of metaphysical inquiry is going to be very helpful here since it is very much a kind of 'owl of Minerva' approach. We can reconstruct the 'absolute presuppositions' of past thought once it is really past, but not those of our own thought-epoch since we are immersed in it. The most intense debates and contestations over concepts, categories, criteria and the like would affect only the explicit or explicable 'relative presuppositions', at least if we take Collingwood's approach as it appears in his texts without seeking a mitigation of the claims to render it more serviceable in the confrontation of rival visions of reality.

4 *De jure* is a Latin phrase that means 'according to law'. It can be contrasted with the Latin phrase *de facto*, which refers to something that exists 'in fact', even if it is not prescribed by the law.

5 A term popularized by Michael Oakeshott (1959).

6 This point hardly needs emphasis, but one might recall his famous lecture, 'Two Concepts of Liberty' and see a good commentary such as the one by John Gray (1963).

7 For a range of points of view on the issue of incommensurability in moral philosophy see the recent collection edited by Ruth Chang (1998).

8 *La rage de vouloir conclure* is a French phrase, stated by Gustave Flaubert, that roughly translates to the 'obsession to draw a conclusion'.

9 This expression is not now widely used, but it was quite frequent in the 1950s and 1960s in the United States when a variety of techniques and disciplines like operations research, the Theory of Games, cybernetics, systems theory, statistical decision theory, econometric modelling, and so forth were supposed to be the sufficient instruments for scientific policymaking, for scientific prediction and control, in the age of the Cold War and the National Security State. It had some impact on urban affairs and business management in the bigger corporations, but the political climax of this movement, of which the Rand Corporation and the Hudson Institute were paragons, was perhaps the reform of the Department of Defense of the United States by Robert McNamara (who at the time of his appointment had been chairman of the Ford Motor Company), though the role of game-theoretic models of strategic interaction continued to be important in strategic studies, above all in the area of nuclear-weapons control. However, there is still a journal under the title, though there is comparatively little talk about the idea of a policy science. A parallel development took place in the USSR, but the total scientific management of society, which was envisaged under the leadership of cyberneticians as an extension of defence management was aborted by the political leadership, who feared loss of control and privilege, and never spilled over beyond the war establishment (and its core, the nuclear-weapons systems).

10 In a recent interview, Mary Warnock (aka Lady Warnock), who has engaged with bioethical issues for the last twenty years or so, declared that she could not see any valid, substantial

arguments against *any* of the genetic engineering projects concerning human life that have been discussed in recent times (2002). She had serious misgivings only about the prospects of bioengineered immortality but, even in this extreme case, she thought her own misgivings were far more emotive than rationally founded, so that, for this bioethicist, in the end, one might say, to borrow Feyerabend's famous dictum regarding the methodology of science, 'anything goes'. Regarding HRC, she declared it perfectly all right – and yet, oddly enough, went on to stipulate that it should only be allowed after proper consideration, case by case, by a kind of tribunal; amongst other things this would probably precipitate a vexatious waiting list. But this does not make sense: if no substantial objections of any sort can be levelled against HRC, then the presumption must be, in a free society, that it should be allowed without subjecting people to any kind of investigation. If it has to be licensed, what are the objections underpinning the licensing? It sounds like a kind of residual paternalism, if one may put it that way time, so wholly at variance with the times and her other professions. After all reproductive freedom, *freedom of procreation*, freedom to procreate or not to procreate in any number at any time with anyone, within and outside marriage or any other social relationship, a socio-legal condition that even in an earlier and more limited version almost reduced to despair the classical eugenicists, concerned as they were with quality control and prevention of the reproduction of lesser beings, though they continued to seek prohibitions of marriage and procreation regarding the genetically defective, the mentally defective, and those with low IQs, remains largely unimpeded in the West, which has not succeeded in combating low and falling birth-rates. Freedom of procreation is wider than, but stemmed initially from, the near-absolute freedom of marriage, barring age minima, sex-asymmetry – but even that will disappear soon – and the closest blood relationships, for the old Catholic medieval prohibitions of marriage between those related by blood or affinity have otherwise practically all disappeared from civil codes; monogamy is maintained though in fact subverted by serial monogamy or serial polygamy. However, an exception seems to be made for prisoners for, although they are not barred from marrying, according to a recent US Supreme Court decision, they are barred not only from conjugal visits but from impregnating their wives via artificial insemination. To be sure, proposals for the compulsory sterilization, particularly of males, of certain social categories (criminal or psychiatric), not to mention permanent confinement, or passive or active euthanasia, measures endemic in the history of twentieth-century eugenic agendas and, in fact, an officially condoned practice, which in small numbers did not end till the 1970s in Scandinavian countries, continue to be advanced from time to time in Western countries, and not only by manifest crackpots. Oddly in some countries like the UK, procreation rewards like child allowances and family allowances or tax advantages for married couples, which were partly designed for this purpose, have all been under reconsideration. It is also true that to correct the socially undesirable macro-outcomes of this regime of *reproductive laissez-faire*, ingenious mixtures of coercion and market procedures are to be used, instead of the rather brutal coercive measures that have been so characteristic of classical eugenics for negative eugenics at least. A number of economists subscribed to eugenic tenets till the end of the 1930s at least, and have suggested from time to time, notably the proposal by the American economist Kenneth Boulding (1964) for the issue of a fixed number of procreation licenses by the state, which could, however, be bought and sold freely by the original license-grantees. In this way, couples who did not want children or wanted fewer than the statutory maximum, could sell their licenses to those couples who wanted more than the normal entitlement of progeny, instead of a strict uniform maximum enforced with respect to every potentially procreative couple, taking 'zero population growth' (ZPG) as a macro-target. This was a proposal initially formulated at a time when the population question was very much in the air in the United States in the economic crisis period of the 1970s, the time of the first and greatest 'oil shock'. This was indeed basically the same kind of idea as that of the market for pollution licenses for states in the European Union and elsewhere, instituted twenty years or so after Boulding's proposal, and clearly this is a very plausible market proposal for economists in place of purely coercive-administrative measures. The economist Robert Ayres has envisaged the issue of carbon licenses to individuals within a state, and this could be extended in somewhat utopian fashion to the entire planet, which would reduce carbon emissions overall and transfer income from heavy users to light users (Ayres and Walter 1991). Of course, it could be argued that progeny licenses are also licenses to pollute and thus perhaps the same mechanisms should be invoked in all such cases. Some southern European countries have achieved something like ZPG over the last decade or so without any conscious policy towards that effect and, generally, it is now claimed that in practically every developed country fertility is below replacement level. Those countries that experience population increase do so owing to net immigration. Admittedly, what one might call the *freedom of disprocreation*, by which I mean not the freedom not to procreate, but the freedom to abort, lawfully, and with medical assistance, within certain boundaries set by national legislation, solely in terms of a woman's will, is also pretty well established, at least in some Western countries (if not all). In the former Communist countries, with the advent – de facto, if not de jure – of something approaching abortion on demand, although it has been, and to some extent continues to be, bitterly contested in some countries, not least the United States. The comprehensive if not universal medicalization and hospitalization of abortion, which has followed the medicalization and hospitalization of birth, rendered the practice, though lawful, and saving the lives and health of women, an obvious target for physical attack. Both freedom of procreation and freedom of disprocreation, as now extant in the West, can be subsumed, I suppose, under Dworkin's locution of 'reproductive autonomy' (1993), a convenient honorific and euphemistic label for current Western practices, trading on the pre-understood positive content of 'autonomy', but failing to acknowledge the tragic facets of the practices, more evident in the much-quoted paradoxical statement of an American political scientist; 'abortion is murder and I am all in favour of it'. In some Communist countries, at any rate in the last decades before the collapse of the Berlin Wall, such as Hungary, the situation was much the same, with registered hospital abortions at times almost matching the number of live births, reaching, according to the official figures, the ratio of over 90 abortions to 100 live births, though as far as I know the practices were never legitimated under this rubric. Yet total freedom of procreation and disprocreation as a societal configuration was probably initiated in any state in modern times in the early days of the Soviet Union, though fairly quickly it was substantially reversed in the 1930s, as was noted with horror by conservatives at the time, little expecting that it would one day become the normal state of affairs in the West without any particular socialist or communist input but through broad movements of opinion and the emancipation of women. The relations of production of the USSR are quite remote but not its relations of reproduction in the 1920s, which were in some ways a prefiguration of ours. But in any case, if HRC is just another reproductive practice, in every significant respect, then freedom of reproduction by cloning should obtain without question, just as it obtains for all the others. If measures of licensing and quasi-judicial controls are introduced with regard to HRC (it seems this author did not stipulate anything concerning animal reproductive cloning) which simply do not apply to any other form of human procreation, even medically assisted forms with the new RTs, then it appears to signal that it is not quite like the other forms of human reproduction, even those involving the new RTs that do not involve submitting the prospective parents, or the prospective genitrix, or the owner or leaser of the host womb, to any kind of formal, quasi-judicial proceedings. On the one hand, this Peer of the Realm is saying that HRC is perfectly unobjectionable (and, moreover, she castigates, in no meek and mild terms, those who are uneasy about it) but, on the other hand, she is implying that somehow it is significantly different, and problematical, to the point of requiring special legal or quasi-legal procedures of a sort not applied to other reproductive practices (adoption is another matter). People in the West reproduce or do not reproduce at will, unlike the People's Republic of China, with its statutory, and it seems really enforced, limit on the number of children per couple noncircumventable via serial monogamy. In this case, it happens without the market palliative for progeny licenses of the kind suggested by Boulding, well in line with other economists' proposals, but it does have the merit of stark simplicity and uniformity, though with two limits, a limit of two children per couple for the countryside and one child for the urban areas, hence the usual expression 'one-child policy', and if HRC is perfectly okay, why should they not clone at will? Set the people free, free to have themselves cloned, if they so wish, as often as they wish if they can afford to! The technical and economic feasibility of such practices is of course another matter altogether. What are the objections to a regime of *cloning freedom*, of *cloning laissez faire*, as part of the *freedom of procreation*, the reproductive laissez faire, which obtains in the West and would not also be objections to (human reproductive) cloning per se? It is not clear to me what they would be.

11 The concept of 'medical utopias' was advanced by the Franco-American biomedical researcher and humane scholar René Dubos, in an article in the journal *Daedalus* in 1959, and subsequently elaborated in his book *Mirage of Health* (1959b). On the vision of 'perfect health', especially in contemporary biomedicine, see the thoughtful work by Lucien Sfez *La santé parfait* (1995), which compares the United States, Japan and France. Outside biomedical circles, other scientists, such as Stephen Hawking, in a paper included in his most recent book, suggest that the species needs urgent improvement, in particular a bigger brain, in order to face environmental and other challenges (Hawking and Mlodinow 2010). This is quite a demand, since the human brain has been fairly constant in size for perhaps hundreds of thousands of years (though the matter is disputed), and such an increase in brain size would require inducing corresponding changes in female reproductive anatomy, most unlikely if not for physical at least for social reasons, unless ectogenesis were contemplated, which might well be the only way for this augmentation to take place. Indeed, this desiderated increase in brain size for our species has been specifically advanced by another recent author as another reason for proceeding towards ectogenesis and bridging the time-gap between IVF and the incubation of the prematurely born and rendering the whole process from fertilization to viable existence at about 9 months entirely in vitro. Of course, neuroscientists now expecting to develop powerful technologies may have something to say on this matter, as on all issues of the mind–brain, including the envisioned megacerebral being. Computer scientists have their own expectations and not only they. For instance, the neuroscientist W. H. Calvin (2008) takes this option or wager on the urgent massive development of AI with a view to reaching the goal of *autonomous* AI as soon as possible so as to enable us to survive the impending large-scale adverse environmental changes. The price to be paid subsequently is either exiting from our biological condition altogether or the supersession of homo sapiens. We have to be rescued in order to be replaced.

In France, recently, public authorities intervened to discontinue, against the wishes of the deceased, the cryogenic perpetuation of a corpse, inspired by the American cryogenics movement ('French Court Rules Against Frozen Couple' 2002). It is not clear whether the Republic fears the vain hopes of terrestrial immortality (though Robespierre professed a belief in the immortality of the soul) or just the non-democratic restriction to the few who can afford it.

12 This obsession with longevity and immortality was marked in Russian pre-revolutionary culture, not least in some strands of Orthodox religiosity (the cult of the *starets*). The embalming of Lenin was partly motivated by the Marxist-immortalist hope, entertained by Bolsheviks or fellow-travellers of the Bolshevik Party influenced by the thought of a very remarkable visionary of technology, N. Fedorov, that under socialism, rapid scientific progress would render possible his resurrection though obviously the final decision was Stalin's and that exclusively for mundane, and very shrewd political reasons, as was the embalming of some other top Communist leaders in other countries (never more than one per country). The embalming was done with the assistance of German doctors; the continuance of German expertise and interest in the matter emerged in November 2002, when it was reported that the brains of several members

of the Baader-Meinhof gang had been embalmed secretly and preserved for examination, to ascertain the features responsible for their criminal behaviour. They did find abnormalities in the brain of U. Meinhof, who had already undergone brain surgery some time before she turned to terrorism. Now, with increased longevity and the prospect of 'regenerative medicine', being a centenarian will cease to be unusual, and some science writers are even beckoning with the prospect of two to three hundred years as a perfectly attainable lifespan (Wade 2001).

13 *Verstehen* is a German term that means 'to understand'.

14 As argued by the French scholar Pascal Bruckner in a recent book (2000).

15 *Fait accompli* is a French phrase that means 'accomplished fact', or something that has already been done.

16 *Sui generis* is a Latin phrase that means 'of its own kind'.

17 Cf., David Lamb's *Down the Slippery Slope: Arguing in Applied Ethics* (1988). This is the best text and the only book exclusively on the topic. It has been much referred to in texts on fallacies, informal logic and rhetoric.

18 The historical semantics of the locution 'social engineering' over at least the last sixty years or so, deserves a special study. Popper's strictures in *The Poverty of Historicism* (1957), it will be recalled, were against holistic social engineering, not against social engineering as such, for he was a keen advocate of what he called 'piecemeal social engineering' and democratic social reform. Indeed, he is still being criticized in print by some conservative and right-wing scholars for his social democratic predilections, for his stress on the need for social protection of the weak, associated with his faith in the latter. It is quite wrong to conflate Hayek's and Popper's normative stances on the role of the state in democratic capitalism, for the latter was much more positive; regrettably, this is quite a widespread error.

19 Derived from Latin and ancient Greek, *ecumene* refers to the inhabited world.

20 *Interlocuteurs valables* is a French phrase the means 'valid speakers'.

21 The Stoic term, 'cosmopolis' was recently revived by the philosopher S. E. Toulmin, who has also engaged extensively with bioethics, as the title of a book (1990). The Stoics were the first to envision a utopian polity on a world scale instead of the island or city-state utopias which were the dominant and indeed almost exclusive form of utopianism in the classical world, and even in the modern world until the nineteenth century. Some of those who speculated about the emergence of a 'world-brain' six decades ago also posited the emergence of a 'world-state' to go with it, as was the case with H. G. Wells.

22 On the 'principle of permanent control', see Felix Kaufmann's *Methodology of the Social Sciences* (1944); though he did not refer particularly to Popper in this context, he had been quite familiar with Popper's work in Vienna. The locution 'permanent control' has the advantage of not carrying the load of implications associated with specific methodologies of science or epistemologies such as 'falsifiability', and encompasses both falsifiability and criticizability in Popperian terms.

23 A thesis advanced by Simmel, notably in the text translated decades ago as 'the web [sic] of group affiliations' and restated by C. Bouglé in *Les idées égalitaires* (1899).

24 We have coined this neologism, *Homo heterarchicus*, in order, in part, to pay homage to the psychoneurologist Warren McCulloch, a key pioneer of the Cybernetic Age, for it was he who advanced the concept of 'heterarchy' in his famous studies of neural nets in the 1940s, but also in part to provide a counterpoint to some key macro-sociological concepts advanced in the comparative study of civilizations. The concept of *homo hierarchicus* was formulated by the French anthropologist Louis Dumont in his major studies of the Hindu caste system and utilized by him in his studies of a number of major currents of European thought (the impact of Calvinism and of German Romanticism). His book on the social ontological presuppositions of Western individualism was entitled in the French original, *Homo aequalis* (1977). The English translation of this book uses a different title and therefore the allusion to his previous book is lost. Curiously, the modern type of character structure in the West – which David Riesman

delineated as other-directed, succeeding the Calvinist conscience of the inner-directed person (the 'guilt culture') – was partly explicated in terms of a technological metaphor supplied by none other than the theorist of Oriental Despotism, K. Wittfogel, that of radar, which presumably was supposed to be switched on most of the time (the inner-directed, conscience-driven, or self-steering man was depicted in terms of another new-fangled technological metaphor, that of the gyroscope, another, older cybernetic mechanism). From continuous attention to the radar screen and gearing oneself to its indications in the ever-changing presentation and re-presentation of the self in daily life (Goffmanesque Man seems to proceed in this fashion) is not such a leap to being permanently online and living on its terms, involving fuller and more vivid screens (soon to surround us everywhere at all times), and abundant instant unceasing information on all sorts of predicates.

25 'Reverse serendipity': giving as an example the late discovery – or rather late public denunciation and in this sense late public discovery – of the adverse and often disastrous effects of pesticides on untargeted animals and humans after decades of massive and ever-expanding use throughout the world (Dowd 1989, 83). But, similarly, we might, taking Dowd's lead, speak of *reverse moral serendipity* when practices, not necessarily technological, which have been accepted, even taken for granted, are reappraised in ethical terms, in the light of experience and reflection, not least from the experience of disasters, which still seem to be the greatest teacher.

26 Adverted in Rawls's *A Theory of Justice* (1971), it has been developed by Norman Daniels in a number of papers collected in a recent book.

27 The concept of Protean Man (meant to be gender-inclusive of course) was developed by the psychologist Robert J. Lifton in the 1960s. 'Protean Man' is one of several models developed to portray the putative successors to the Protestant ethic and its social character-structure: the 'Maitreyan' value-system (Morris 1964), an increased aesthetic-expressive emphasis in the dominant American value system (Kluckhohn 1951), the 'Freudian ethic' (LaPiere 1959), 'psychological man' (Rieff 1959), 'the organization ethic' (Whyte 1956), 'the other-directed' character structure (Riesman, Glazer, and Denney 1950), 'self-actualizing man' (Maslow 1954). The latter has been a particularly influential model, even in the management literature and business school textbooks, and has also had a significant impact in the comparative political science of advanced industrial societies, in the concept of 'post-materialism' of R. Inglehart (1977) as an increasingly more prominent phase of value-concerns in the richer societies when the quality of life becomes more important than material well-being though environmental issues are often seen as issues of public health as much as aesthetic concerns.

28 For Protestantism and its campaign against natural (or 'spiritual') magic see Keith Thomas's *Religion and the Decline of Magic* (1971); of course, the obverse of this obsession was the extraordinary historic wave called the 'witchcraft craze', and the consequent burning of witches of the sixteenth and seventeenth centuries, pursued with much more vigour in Protestant countries and territories than in Catholic ones, where instead Jews and heretics were burnt more often by the Inquisition.

29 In some ways, given the importance of utilitarian prepossessions in our civilization, it is surprising that sentient animals, perfectly capable of pain, by the very definition of sentiency, have not received more protection for the Cartesian denial of sentiency to animals, a portion of common sense or folk psychology to which Descartes took particular exception in his attack on prejudices – though one could think of far worse ones prevalent in his time. The Cartesian doctrine that animals are but natural automata, has not reached as deeply as other Cartesian myths such as 'the ghost in the machine'. This Cartesian position regarding animals as insentient, however, despite appearances and common-sense belief, still enjoys ardent defenders, amongst some contemporary philosophers at least, a sign of the polarization in this field. One of the leading contemporary advocates of protection for animals, or at least for the sentient ones, is the utilitarian ethicist Peter Singer, but it is curious that such utilitarian advocacy has been so late in coming, or in receiving widespread public notice. Of course, the defence of

animal welfare and rights can be legitimated in terms of other kinds of ethical doctrines. No utilitarian can defend animal *rights*, in the strict sense of the term, for this is not a legitimate ethical concept for utilitarianism, though they may speak of animal 'interests'.

30 In recent years the 'Flynn effect' has been reported. This effect was inferred from a variety of reports from Western and Asian tiger countries, which pointed to a steady increase in measured intelligence over the last three decades; the more people, or at least young people, watch television, the cleverer they get! In November 2002, another report was issued by some American psychologists (Ulrich Neisser et al.) claiming a 25 per cent rise in IQ in three generations, that is, claiming that today's children are 25 per cent 'cleverer' than their grandparents overall, though not in terms of verbal intelligence, which has stayed put, to perhaps no one's surprise. This rise in measured intelligence is supposed to be caused by three key factors: improved nutrition, computer games, and Internet surfing (computer games particularly are supposed to exercise the brain to no end!). Still, other side effects of TV, computer games and electric lights have been detected; children are significantly more afraid of the dark than those of the previous generation. More broadly, psychologists have argued that the rarity of the experience of total darkness in the West today could be stifling the imagination: 'Being in the dark gives them a blank canvas on which to paint their own imaginary pictures. It can be very stimulating for them to play and entertain themselves in the dark, because all the images they produce will be unique, but this rarely happens now' (Dr. Aric Stigman quoted in the *Times* 2002). Even fear of the dark, according to the same report, seems greater than before (I am not sure with what confidence one can say the it is 'greater than ever before'), but is now focused on 'ready-made images imprinted on children's minds from TV, cinema and computer games'. Anyway, *mehr Licht* has not turned out to be an unqualified blessing for human development. However debatable this 'Flynn effect' may be, at least this reported IQ rise or inflation has stopped geneticists' concern over the issue of diminishing (measured) intelligence that so alarmed them in the 1920s and 1930s, a belief that continued to be reiterated well into the 1960s, if not over the accumulation of deleterious mutations as a whole. The history of 'measured intelligence', so important in practical terms in the framing of educational policies and school systems, does seem an extraordinary, monumental history of mis-measures, as S. J. Gould put it and investigated in a striking work. For many decades, IQ measurement has notoriously played its part in sustaining biological racism, most of all in the specific case of African Americans. In addition to being cleverer, we are also taller and fatter. Despite the cult of physical fitness, it is not clear whether the median male is fitter today than a generation or two ago. But we are engaged in the cultivation of (male) athletes where records in a number of areas have been rising steadily over the last forty or fifty years, though perhaps now reaching physical limits, or in the jargon 'maxed out humans' physical potentials'. The threat of undetectable nature-identical performance-enhancing drugs and of capability-enhancing 'gene-doping' (to induce the production of hormones or other performance-enhancing substances) or a kind of genetic enhancement of athleticism, seem very real (one wonders why similar drugs or 'gene doping' are not now widely available for those having to perform in competitive examinations, interviews, tests of various sorts). Female athletes certainly have not yet reached the limit of their capabilities, and the same goes for amputees, the disabled, the aged. And there are less-intrusive, non-genetic, non-chemical analytical technologies to aid performance such as high-speed videography in the genealogy of Etienne-Jules Marey, and near-simultaneous feedbacks. Others rely on guided imagery and neuro-emotive techniques as an alternative to genes and drugs.

31 This appears in the second volume of his posthumously collected papers (Hamilton 2002). H. J. Muller put this point as the 'ratchet effect' of deleterious mutations. Having discovered the mutagenic effects of ionizing radiation in laboratory experiments on *Drosophila* already in the 1920s, he campaigned for many years against above-ground atomic tests, arguing that they were increasing the load of these mutations unnecessarily and dangerously, which partly explains the title of his biography, *Genes, Radiation and Society: The Life and Work* of H. J. Muller

(1981) by E. A. Carlson: a campaign that was eventually successful, with most subsequent atomic bomb tests – though not all – being underground, and raised awareness about the mutagenic and carcinogenic effects of new technologies, as in the case of synthetic chemical pesticides which also eventually led to the limitation of pesticide use and legal and regulatory controls. But he also believed in the dysgenic over-all trend, at least in Western societies, in the absence of corrective interventions by genetic counselling and appropriate public policies. Francis Crick, the co-discoverer of the double-helix model of DNA, has also gone on record recently to express concern over dysgenic trends, and has argued that there is absolutely no right to life at birth for those who are not genetically sound in every way (presumably in addition to the criteria already applied in the determination of the kinds of foetal euthanasia already extensively practised in Western hospitals, as also, if to a much lesser degree, neonate euthanasia). Biological panics, specifically dysgenic alarms, occur from time to time, and a much-publicized recent one, which continues to appear in the newspapers, has been the concern over the alleged falling sperm counts amongst Western males over the last ten to fifteen years and, more generally, rising male infertility, though both the facts and the putative causes are in dispute; the estrogenic effects of chemical pesticides on animals as well as humans have been ascertained, to add to the mutagenic, teratogenic, carcinogenic, genotoxic and other adverse biological effects previously determined, not to mention the secondary lethal impact of various types of biocides on non-targeted mammals, birds and other animals. The widely read popularizer, geneticist Steven Jones, has argued that the Y chromosome responsible for male sexual characteristics is doomed, though its disappearance may take quite a number of generations yet (2003); still the trend according to this author is unmistakable, so inborn maleness will be over and done with, though surely much else that we prize, including the world, may come to an end before such a state of affairs is reached.

32 *Faute de mieux* is a French phrase that means 'for lack of something better'.

33 'Thou shall not kill, but thou shall not strive officiously to keep alive'.

34 I specifically refer to Austria, owing to the great medical tradition of the country, Vienna being one of the leading capitals of Western medicine, and the quality of its biomedical thought (though it had also been a great centre of 'medical nihilism'). Calls for compulsory sterilization of certain social categories defined in psychiatric or legal terms (particularly of males) have been made in recent years by at least some minor public figures in the West, some with ministerial experience, partly on the fiscal grounds that the welfare state cannot cope with the expenses accruing from the care of their progeny; this fiscal argument was also a standard Nazi argument to justify the killing of thousands of the incurable, of hopeless invalids, of the senile, and others who were a permanent burden on the state. It has been alleged recently that something like a quarter of a million Indians were coercively sterilized in Peru under the two-term presidency of Fujimori, which, if substantially true, would probably constitute the largest population in any one country to be subjected to this since 1945. Calls for the compulsory castration (at least the chemical kind) of recidivist rapists have also been made in recent years in Western countries in a variety of quarters. Italian government ministers have recently called for the castration of rapists amongst immigrants. The sterilization and elimination of mixed-race persons in the then-German colony of South West Africa was suggested by the leading German eugenicist, Eugen Fischer – the mixed race person being a special horror for German eugenics – and it is not altogether impossible that drastic negative-eugenic policies, turning into a kind of world laboratory of state negative eugenics, would have been carried out in this German colony if World War I had not intervened; (it will be recalled that eugenic legislation started in the West in 1907). In fact, persons of mixed African and European descent in the Reich (particularly children of French Africans and German women during the period of French occupation of the Rhineland, in which French African troops were stationed) were subject to compulsory sterilization after 1933 within the same policy complex that also targeted the incurable and the mentally ill.

35 Even Keynes subscribed, at least for part of his life, to a eugenic creed, with clear racist views, and sympathized with coercive negative eugenics, though he published little directly on these themes. This unpleasant matter has been very fully discussed with extensive examination of unpublished texts in the recent book by John Toye (2000). No doubt some will see a deep connexion between this stance and his other more public doctrines on economic policy, though this would seem to me far-fetched and lapsing into a clear-cut form of 'the genetic fallacy'.

The German sociologist Theodor Geiger, who sympathized with the Social Democrats, was a staunch supporter of eugenics, but he left Nazi Germany right at the start, and was in no way compromised with the regime. Curiously, his writings on the subject went completely unmentioned after 1945, when he was enthusiastically welcomed back in a democratic West Germany and have only recently been referred to. He was a sociologist of some distinction, and was honoured with a volume, edited by Renate Mayntz (Geiger 1969).

Among the most enthusiastic and vocal were the Austrian C. Ehrenfels, a key figure in the development of the theory of Gestalt qualities and of axiology, and the British pragmatist or humanist F. C. S. Schiller (who lived in California after Oxford), both rather ambiguous about democracy. Among those who expressed at least sympathy with the eugenic outlook, without being obsessed with it, were also philosophers of liberal-democratic persuasion like Léon Brunschvicg, the leading academic philosopher in France between the two world wars, who praised Plato for his eugenic and feminist proposals in *The Republic*; the young Raymond Ruyer, later a liberal thinker and a major analyst of utopianism in postwar France, and the American W. P. Montague, who praised eugenics as a movement committed to the future like the pragmatist movement as a whole, and incorporated it in his 'Promethean religion' of human emancipation and world conquest. Brunschvicg exemplifies what may be called the eugenic-feminist syndrome, not always accompanied by racism, as it was not in this case; his wife became the first ever woman member of a French government, as undersecretary of state in the Popular Front administration. Although women did not have the vote in France, the leader of the government and of the Socialist Party, Léon Blum, had promised to appoint a woman to a ministerial post. The subject of the attitudes of philosophers to eugenics has not attracted much attention, except in the case of Nazi Germany. Heidegger's borrowing from and friendship with the arch-theorist of German racism, the anthropologist and long-standing leading eugenicist, Eugen Fischer, has been recently analysed by various scholars in the light of Volume 16 of Heidegger's *Gesamtausgabe*, and it is quite clear that he took over biological racist discourse without circumspection in various speeches in 1933 (Münster 2002). Natalie Harris Bluestone does refer to a Nazi attempt to enlist Plato in their canon but does not raise the broader issue of eugenic engineering of society (1987).

Another example is Winston Churchill at the first international eugenics congress. By contrast, the address to this congress by the philosopher and Conservative politician A. J. Balfour was remarkably thoughtful and cautious.

36 Especially in the substantial work of the otherwise very respectable British geneticist C. D. Darlington (1969). The times were not ripe, and his work did not elicit a sympathetic reception at the time, but it would be interesting to do a systematic comparison of his works of genetic-biological grand historical sociology with those of later tillers of the same soil, the self-named sociobiologists (who do not refer to it). The obsession with the decay of civilizations (with the theme of degeneration) has been very marked in modern biological racism since Gobineau and, if with less eloquence perhaps, in the writings of eugenicists since the founder's *Hereditary Genius* (Galton 1869).

37 The American geneticist H. J. Muller was working on similar lines and wrote an essay on this theme in 1925, which he did not publish at the time, but it appeared ten years later in *Out of the Night: A Biologist's View of the Future* (1935), which was much praised by Haldane, who had also become a Marxist by that time. The word 'future' had become mandatory. Changes in

reproductive institutions had been taking place in the West, especially in Protestant countries with the gradual disappearance in practice and in law of the vast array of prohibitions on the marriage of those related by blood and affinity according to the criteria laid down by the medieval Catholic Church, partly preserved by the Anglican Church, some like the prohibition of the marriage between a widower and his sister-in-law into the twentieth century (Goody 1983). The upper-middle classes in England could not have managed without the supersession of some of these prohibitions. Darwin, after all, married a first cousin (a Wedgwood), and the cousin marriages that tied together the lineages of the Darwins, the Wedgwoods, the Galtons, the Huxleys, the Stracheys, the Keynes, the Haldanes, the Hodgkins, the Cornfords, the Bells, and so forth ran through an elite of professional families. This trend has been analysed in Noel Annan's classic study of the English upper-middle class intelligentsia (1955). They comprised a number of supporters of eugenics, such as Darwin's son Leonard (a key person in the Eugenics Society); Huxley's grandson Julian, J. B. S. Haldane; his sister (the writer Naomi Mitchison); J. M. Keynes, amongst others; more recently, David Galton published a book on eugenic lines (2002). On the anti-eugenicist side, there is at least one Wedgwood who opposed a eugenic bill in the House of Commons in 1912.

38 In fact, of all those 'future' books of the 1920s, it is the only one that has received recently the accolade of a collection of essays by contemporary scientists.

39 More precisely, by the *academic* scientific community, for plant breeders in the United States were more receptive – after all Mendel's classical experiments were with plants – if more eclectic. The so-called rediscovery of Mendel in 1900 was also according to some accounts a reinterpretation of Mendel's work as ascertaining the laws of heredity.

40 An interesting figure: in addition to his major contributions to statistics and the history of statistics, a supporter of the equality of women and men, a socialist of sorts (certainty wanting the betterment of the conditions of the great mass of the British people), but also an imperialist, if a Social Imperialist (imperialism for the economic benefit of the masses), and a racist in the sense of being a believer in the natural overall superiority of some races and determining their treatment accordingly. He lived to praise Hitler and his 'eugenic' measures publicly (as did Shaw, though privately).

41 See Charlotte Haldane's *Man's World* (1926) for the original edition of her science-fiction novel.

42 It will be recalled that Icarus, who came to a sorry end, was the son of Daedalus.

The book in question was *The Scientific Outlook*, originally published in 1931, a harsh critique of scientific projects of large scale, thoroughgoing societal transformation through industrial, biotechnological, and psychological techniques or some combination thereof, and to some extent a devastating critique of contemporary science itself, not only as compromised in these specific projects, but more widely, inasmuch as it may involve 'power-thought' or 'power-knowledge' (Russell 2009). The very expressions he coined, or at least used, long before the immensely quoted Foucault, not to mention critical philosophers of technology, rather than knowledge for the sake of truth, were out of the love of the world. In the preface to the first post-war reprint of the book, in 1949, Russell noted the similarity of themes in Huxley's novel and his own book, coincidentally published for the first time in the same year. This work was reprinted again, in paperback, this year, as part of a project of reprinting a number of the works of the author. The flavour of Russell's argument may be conveyed by the following quote: 'Science in its beginnings was due to men who were in love with the world. (…) They were men of Titanic passionate intellect, and from the intensity of their intellectual passion the whole movement of the modern world has sprung. But step by step, as science has developed, the impulse of love which gave it birth has been increasingly thwarted, while the impulse of power, which was at first a mere camp-follower, has gradually usurped command in virtue of its unforeseen success. (…) Thus science has more and more substituted power-knowledge for love-knowledge, and as this substitution becomes completed science tends more and more to become sadistic' (Russell 2009, 197–98).

43 It might be called 'horizontal genetic Fordism' being the production of a multitude of genetically identical human beings at a given time. The techniques of HRC would enable a kind of potential 'vertical genetic Fordism', the production of any number of genetic copies of a given prototype over time in successive generations.

44 Dolly, a female domestic sheep born in 1996, was the first mammal cloned through nuclear transfer.

45 The Marxist terminology is not accidental here, for Haldane was formally converted to Marxism in 1936, and was for many years a member of the British Communist Party. He published many articles of scientific popularization (of which art he was a master), in the British communist party's daily newspaper, the *Daily Worker*, and wrote a book on the Marxist (dialectical-materialist) philosophy of the sciences. The geneticist and staunch eugenicist, H. J. Muller, an American who took his Marxism seriously enough to leave the United States and go and work in the Soviet Union for a number of years, used this terminology of forces and relations of reproduction. Even earlier non-Marxist socialists agreed that given that social control of production was a good thing, so also would social control of reproduction, according to the biologist L. Hogben, writing in 1919 (Mackenzie 1981, 460). Regarding the separation of sex from reproduction and vice-versa, Baudrillard commented somewhere that we have moved from situation in the 1960s when the maximum of sex with the minimum of reproduction was all the rage, to a situation in which we seek to combine the minimum of sex with the maximum of reproduction. The very abolition of sex, half-entertained by some utopians of the eighteenth and nineteenth centuries, Godwin, Condorcet, Comte and others, and more recently by a messianic visionary of technology such as Fedorov, as a follow-up to the termination of reproductive sex (and the institution of non-sexual reproduction), has reappeared in the scenarios and the agendas of some sociotechnical cyberneticians: the abolition of sexual intercourse as wholly unnecessary, of non-genital sexuality as well (presumably including sexual caresses), and of sex-differences. Instead cyber-sex will take over. Manicheans, Gnostics, Doukhobors, Shakers, and so forth would have rejoiced, though amazed that this should come about in such an irreligious, hedonistic and materialist culture. Maybe the sexual revolution of the late 1960s and its reflexes today will eventually be seen as just a final civilizational paroxysm before the unforeseen obsolescence of sex.

46 Haldane was delighted when his third wife, the biologist Helen Spurway, worked on parthenogenesis; this topic has been an obsession of anticlerical biologists and science popularizers.

47 It has been reported that a Texas firm (in fact the company alluded to in the text) has actually succeeded in cloning a pet dog, for the fee of a few thousand dollars, though I have been unable to check this report. Nevertheless, the cloning of pets, like the cloning of livestock in general, has not yet entered mass production, and has not impinged much on the collective memory, unlike Dolly.

48 I have tried to take into account the developments that have come to the fore since Troy Duster's book *Backdoor to Eugenics* (1990), perhaps the first study by a sociologist of the new eugenics; this book was translated into French, with a preface by Pierre Bourdieu. No comparable token of interest has appeared in the UK.

49 *Tout court* is a French phrase that means 'very short', 'simply', or 'without anything else'.

50 Nozick advocated a kind of strong culture of rights. Rights are invulnerable, so presumably the rights to take part in what he called the 'genetic supermarket' are also invulnerable.

51 A term invented, I believe, by the sociologist Evelyn Katz Rothman.

52 A leading American geneticist taking part in a Congress in Rome in the 1930s took the opportunity to write to Mussolini about the desirability of eugenic reform. Mussolini had already expressed sympathy for eugenic concerns, though under his rule the chief priority was to increase fertility. Another, eventually a Nobel Prize winner (H. J. Muller), then living in the Soviet Union, wrote a letter to Stalin. Hitler did not wait to be written to, though in any case, as Führer, he was congratulated by some American geneticists/eugenicists for his eugenic

wisdom and courage (most of his programs were negative-eugenic) for his eugenic programs, which started shortly after the take-over in 1933. Muller's letter to Stalin urged him to set in motion a vast programme of positive eugenics to encourage the reproduction of superior people, of the best, now that class privileges and social inequality no longer obfuscated true inborn superiority, and advance the progress of socialism biologically, increasing substantially its comparative advantage over capitalism in a distinctive way, which was barred to capitalism, he thought, by its very nature. He did not get a reply, if indeed Stalin ever became acquainted with the gist of it. After Muller returned to the United States, he was instrumental in setting up perhaps the first sperm bank, the donors being Nobel Prize winners (as he was), and the like, so capitalism did prove somewhat receptive to his plans to encourage the genetic *upgrading* of the country even if in a very 'micro' way. It was however a far cry from this sperm bank, and even all the others that have been set up in the United States since then put together, and his extraordinary vision of further improving a socialist society eugenically (which he had regarded in the 1930s as a necessary starting point and the only sound social and economic basis for his biological programme) to secure 'genius' level for practically *everybody*. Moreover, this could be accomplished within a few generations – and that, he stressed, would not be the end-state by any means. This vision was set forth in his book *Out of the Night: A Biologist's View of the Future* (1935); Muller had recently been living in the USSR. The book was published by the Left Book Club, at the time the major book-publishing enterprise of the Left in Britain, most of whose publications were directly political ones on current issues, and it was probably read by tens of thousands of people. It was obviously seen as a kind of Left Eugenics – and in the blurb his fellow geneticist and at the time also fellow Marxist, J. B. S. Haldane, stated that if Muller's proposals were adopted, the results would be as important as those of the Industrial Revolution. I think that he, as a pro-Soviet socialist, also meant to say, or to imply, that they would be almost entirely to the good, and without the terrible suffering that accompanied its predecessor, being scientifically planned by benevolent and far-sighted scientists and policymakers in a top-down fashion.

53 Another American, geneticist and Nobel Laureate, Joshua Lederberg, had earlier compared modern genetics and its technological capacities to alchemy, and even coined the word 'algeny' in a very positive spirit, a word used subsequently by Jeremy Rifkin as the title of one of his books criticizing technological abuses (1983).

54 The philosopher and historian of thought, René Berthelot, in a series of 1930s articles in the *Revue de Métaphysique et de Morale* (unfortunately never collected in book form) presented the history of this strand of thought. The revival of astrology, as of other forms of natural magic in the fifteenth and sixteenth centuries, was by no means wholly adverse to science. Note that while genethliac astrology was concerned most directly with the life chances of individuals, astrobiology, as characterized by R. Berthelot, was most concerned with the fate of kingdoms and empires, of collectives. Curiously, another biologist, S. J. Gould, also wrote that our fate hinged literally on the stars, in referring to the rise of the mammals, and in particular of our species, as a result of the contingent event of the last great extinction of species on earth (prior to the ongoing one), the fifth great extinction caused by meteoric impacts. Of course, there are still those who have reanalysed astrological hypotheses and found them testable and corroborated, notably Maurice Gauquelin, too serious a scientist to be ridiculed, too heretical not to be ignored.

55 The Australian philosopher Douglas Gasking later argued for a similar view, and, in a more qualified way G. H. von Wright did so as well. The term 'interventionism' as contrasted to strong or hard determinism, was suggested by Lewis Feuer in his critique of historical determinism, especially of the Marxist variety.

56 It is remarkable how little attention social scientists pay to envy, at least under that simple monosyllabic name, even though they are very fond of monosyllabic mantras like 'risk' and 'trust'. Yet the inventor of the concept of 'externalities' in economics, and the author of the first systematic text on 'welfare economics', at least the first under that rubric, A. Pigou,

focusing in effect on negative ones, gave envy as a prime example of the negative externalities of consumption (1920). Other economists might well argue in a neo-Mandevillean fashion that it is a positive externality since it fuels consumption and increases the search for gainful work; liberals tend not to use the word but rather a variety of euphemisms. The only sociological treatise on envy, at any rate of any sophistication, appears to be that of the conservative sociologist Helmut Schoeck (1969).

57 Reprogenetics is a term coined by the geneticist Lee Silver, an enthusiastic advocate thereof.

58 Per the existentialism of renowned French philosopher, novelist, and literary critic, Jean-Paul Sartre (1905–1980), a person's existence is prior to their essence. This claim ultimately refutes determinism and upholds the notion that people can define themselves (Sartre 1946).

59 According to recent reports, a number of men in the UK and the United States have undergone plastic surgery (as well as a number of complementary practices like tattooing, skin treatments, implants, dental work to acquire fangs, tongue-forking, and so forth) over a number of years and hundreds of hours of operations to turn themselves into leopard-like, lizard-like, cat-like figures, through identification with animals, a kind of euphenic totemism. It does show how the freedom to reinvent oneself in any and every way (a realistic freedom if they can afford it) has taken some people down a somatic modification path. Genetic somatic modification is probably still too expensive and too regulated for them; so far, it seems only males have pursued this path, but it may be just a matter of time before women do the same. Perhaps another Dworkin will come along and sum it all up in terms of a value of 'somatic' or perhaps better 'transsomatic autonomy', the right to one's somatic modification, by all surgical means, even the right to look nonhuman, to be added to the 'reproductive autonomy' that has come to play such a central role and to be esteemed so highly in the West, even as integral to a democratic culture. Although practiced by zoophiles, who aim to turn themselves into half-beast half-human beings, it is not very different in its way from the strivings of some mechanophiles, computer scientists who want to become more machine-like themselves, through the insertion of more and more microchips in their own bodies, to keep as long as possible, to play a greater and greater role in their lives. One of them has published a book significantly entitled *I, Cyborg* (Warwick 2002). Soon we will see books or other publications entitled *We Cyborgs* or *Everyday Life of a Cyborg Family*, and so forth. All these developments can be described, of course, as experimenting with the boundaries of the human. It is no longer imaginative literature that accomplishes this, but real techno-biological life in the wake of science fiction, which has a way of becoming factual before too long. As a biophysicist wrote in the 1960s, 'we live in the science fiction of yesterday', which is not all that different from Berdyaev's saying, quoted as an epigraph in *Brave New World* (Huxley 1931) that the terrible thing about our time is that utopias are being realized.

60 Frank Knight, though a Chicago economist, often insisted on the mutability or fluidity of preferences ('tastes', 'desires', 'wants') against the grain of the profession, and used to say that the point is not to implement, 'to reveal' one's untutored preferences, but to develop better and better preferences. The philosopher D. Collingridge, almost as if in answer to this plea, has worked out a 'critical theory of preferences', combining an Austrian perspective with the critical rationalism of Popper. This theory ought to be better known. I refer to it in the second part of my two-part paper on 'Risk, Uncertainty and Eschatology', in *Episteme-Revista da Universidade Técnica de Lisboa* (Martins 1998).

61 On the history and critique of the idea of human perfectibility in the West see John Passmore's *The Perfectibility of Man* (1970).

62 Leon Kass's 'The Wisdom of Repugnance' was originally published in the journal *New Republic* in 1997 and was reprinted in various places, including in *The Human Cloning Debate* (1998b), edited by Glenn McGee, and in *The Ethics of Human Cloning* (1998a).

63 *Tremendum* is a Latin term that means 'awe inspiring'.

64 It has recently come to light that IVF clinics in the UK have used extensively the services of such people as Danish medical students. Apparently, they have generally set a limit of 25 to the number of impregnations per donor, but this has been significantly exceeded in some cases.

We have therefore the surprising emergence in the individualistic contemporary post-modern West of very small family sizes (even with widespread 'serial monogamy' or 'serial polygamy'), of something like *genetic sultanism*, albeit with anonymous donors who perhaps did not indeed intend or expect such a number of impregnations from their contributions. In an age of revolt against patriarchy, feminists, and everyone else, seem to be surprisingly content with its genetic counterpart of *genarchy*. Indeed, some sperm banks were set up by feminist idealists in California to give women greater control over their fertility, in which the same donors – mostly university students – were kept active by these for periods of up to ten years or so, and have thereby impregnated many women all over the world. The stuff can be sent by post for do-it yourself insemination as well as by their partners in the normal way. Indeed, some women clients kept asking for the same donor again and again, a repetitive practice that probably had not been anticipated, but I suppose it falls under freedom of consumer choice or demand-driven IVF since there can be no question whatever of male imposition in such cases. The anonymity barrier of sperm donors to IVF clinics is under severe strain, as recent legal cases in the UK demonstrate, and an apparent change in the climate of opinion (as in the United States, too) also indicates that it may not survive much longer. Even Mary Warnock, who had been a staunch advocate of such anonymity on ethical grounds when otherwise endorsing IVF in almost every shape and form, has now changed her mind on this matter and recognizes the psychological need and moral right of the children of such reproductive practices to know who are their genetic fathers ('their donors' a phrase that makes one think of a 'grant economy'). It seems that in Portugal, and perhaps other countries, it is still believed by lawyers and judges that the anonymity is necessary, desirable and secure, but it is doubtful that the recent experience in the UK will be entirely confined to it. It will be interesting to see how sperm donors will react, not least the unwitting 'genetic sultans' who might come to light under judicially enforced disclosure, taking into account the recently demonstrated pressures by the children of IVF for repealing anonymity in these matters, which have induced former supporters of strict secrecy requirements to change their minds.

65 I have coined the words *genocracy/genocrats* by analogy with the word *biocracy/ biocrats*, terms first used by the British science commentator Gordon Rattray Taylor, author of various books on the biological revolution and on issues of evolutionary theory, in the 1960s. For examples, see his books *The Science of Life* (1967) and *The Biological Time Bomb* (1968).

66 The overlap of the two populations – geneticists and eugenists – was very considerable, and indeed most geneticists were avowed supporters and enthusiastic advocates of eugenics through textbooks, newspaper articles and other channels (Pichot 2000). They even made 'educational films' to propagate their creed, to be shown in schools and to a variety of other audiences. There can be no doubt of their seriousness, and their belief in the necessity, and not least the *urgency*, of the eugenicization of society to obviate national or civilizational disaster.

67 For some interesting comments on this episode in the context of genetics and eugenics in the Soviet Union, see the work of A. Pichot (2000), and see also the account (not cited by Pichot), based on letters of the time sent by Muller to the fellow biologist and fellow eugenicist Julian Huxley, who became the first director-general of UNESCO, as reported in Huxley's *Memories* (1970), where he quotes at length from a contemporary letter to him by Muller.

68 An example is the preface to Shaw's play *On the Rocks: A Political Comedy* (1933), wherein he writes of his support of eugenics.

69 Measures of negative eugenics today tend to be of the kind already adverted, of foetal and neonate euthanasia, as practised in Western hospitals (euthanasia at least in intent). Of course, the bulk of medically assisted abortions do not have anything to do with this.

70 *Cultiver notre jardin*, stated by Voltaire, is a French phrase that means 'cultivate our garden'.

71 The allusion is of course to the title of Leo Marx's famous book *The Machine in the Garden: Technology and the Pastoral Ideal in America* (1964).

Chapter Four

THE BODY VANISHES! MOMENTA OF DISCARNATION IN TECHNOSCIENCE TODAY

The sense of 'body' or rather of 'the body' that prevails today over all others, almost as if it were the 'unmarked' or default sense, is that of *human* body. When books, articles, exhibitions, art catalogues, thematic magazine issues or specialist learned journals (like the British-based *Body and Society*) appear with 'the body' as keywords in the titles, the sense that is taken for granted unless corrected is that the reference is to the human body. There has been quite a massive shift in learned discourse over the last decades, certainly over the last century or so, both in English and in other Western languages, over the incidence of 'body' as (human) body and 'body' requiring one or another of a considerable number of qualifiers not automatically entailing the presumption of human location, although metaphorically the term body is still potent in other co-occurrences (e.g., in the book titles 'Gaia's body', 'Love's body'). In the now increasingly distant past, 'body' was very generally accompanied by other terms, to make up such composite expressions in current use such as 'material bodies', 'natural bodies', 'heavenly bodies', 'celestial bodies', 'the body politic', 'the King's two bodies', 'the body ecclesiastical', 'the world's body', 'body of knowledge', 'body of work', 'bodies of opinion', 'the mystical body of Christ' (not to mention more colloquial expressions such as 'bodies of water', 'bodies of land', and so forth and English expressions, very much alive, originally jurisprudential, drawing on the Latin for body, such as 'corporate', 'incorporated', and so forth). Legal English also retains the Latin 'corpus' in such expressions as – most famously – *habeas corpus*, and *corpus/corpora* appear also in semiology and literary theory.[1]

Even a summary treatment of the historical semantics of 'body' would require a substantial volume, at least as substantial as that published some years ago on the incidence and roles of '*res*' in Western philosophical discourse (Kearney 1994). A radical nominalist such as Hobbes was addicted to the use of the term in multifarious conjunctions, as can be seen in any of his key works, and Leviathan and Behemoth were certainly bodies. To be sure, the human body was for Hobbes, as he explains for instance in *Leviathan* (1651), what the animal body was for Descartes, curiously an indefatigable vivisectionist for many years, unlike Hobbes, namely, a 'natural automaton' made up of such parts as cogs, wheels, springs, and so forth, a body-automaton amongst body-automata. In any case, the human body was thus in standard cultivated speech a body partaking of a kind of universal bodyhood, a body amongst bodies of many different kinds, a body in a world of bodies, human and nonhuman, animate and inanimate, individual and collective, material and immaterial, sacred and profane. These linguistic practices prevailed amongst the formally educated over several centuries without generating epistemic malfunctions of the kind that arises when 'language goes on holiday' (Wittgenstein 1953).

It could be argued that these practices were only appropriate in a civilization in which all things were created by one God, all were *creata*, and even though the human body had a very special status, it was the human soul that was truly unique, for only human souls were immortal in the whole of Creation – except for the deviant Christian views, Origenist and other, of those who professed a belief in the doctrine of *apotacastasis*, which could encompass the salvation, resurrection or restoration of all animate beings, possibly some animals, humans, and even fallen angels and demons.[2] And it was certainly favoured by the heritage of Latin where '*res*' (a term which could enjoy the widest generality as 'thing' can similarly in English), came to play a central role in the philosophical language of Latin Christianity, scholastic and post-scholastic, even after the emergence of European vernaculars as languages of high learning and became ever more important in the print culture and 'print capitalism' of modernity. This point has been nicely put in the following text: 'In common usage, "thing" is the most general ontological term in the English language, ranging over events ("a funny thing happened to me"), people ("you poor old thing"), statements, actions ("what a thing to say"), relationships ("baby, we've got a thing going"), intellectual problems ("The time has come, the walrus said, to speak of many things . . . why the sea is boiling hot and whether pigs have wings") – the list could be extended indefinitely' (Collier 1979, 104–5).

Insofar as there was some familiarity with Greco-Roman Stoic thought, scholars would be acquainted with a world outlook for which all was body, or made up, without solution of continuity, of bodies (the 'incorporeals' being purely semantic), a world outlook that has had to be reconstructed from mere fragments. For this corporealist, plenist outlook, the void of the classical Atomists whose thought was revived in the sixteenth century, was never to disappear from the physical sciences. This was despite the onslaughts of the 'energeticists' in the late nineteenth century and their rejection of the atomic or indeed any corpuscularian theory of matter, indeed of the very category of matter as such, as fictional. There was no void in the strict sense but a *plenum*,[3] a continuum of bodiliness, which in turn was revived in field theory in physics from the middle of the nineteenth century. The vacuum of contemporary physics is not a true Democritean void, a sheer absolute emptiness, a nothingness, a meontic rather than ontological category, but a physical reality pervaded by 'quantum fluctuations' (and of course the primary bodies of the Atomists, the atoms, are neither indestructible nor irreducible *discreta*, and the *thema*[4] of discrete entities is anyway under strain in contemporary microphysics).[5]

Oddly enough, contemporary nominalists, physicalists, and materialists largely eschew talk of 'bodies', preferring 'objects' or 'entities', or at any rate even if they would countenance 'bodies', prefer the other terms. Though the logical empiricists at one stage distinguished between the 'thing language' of ordinary language about material things or physical objects and a phenomenalist account of sense-appearances as alternative ways of grounding mathematical–experimental natural science language that need make no reference to either, this usage has lapsed. One exception to this general trend was the work of the Polish philosopher, T. Kotarbinski, a master figure in Polish analytical philosophy, which teemed with logical calculi and invented its own perspicuous logical notation.[6] Kotarbinski advocated 'reism', the doctrine that there are only 'bodies' (as the Stoics also held), a doctrine that could be encapsulated in the ontological maxim: to be, is to be a body. A body is a particular, so that the thesis could be reformulated as: to be, is to be a particular, though it should be remembered that Nature was a 'body' for the Stoics.[7] However, the renewal of pansomatism, even in this guise, has found few followers, though, as we shall see, a 'molar behaviour stance' could be seen as pursuing a similar metaphysical tactic.

Let us stick now, unless otherwise stated, to 'body' as the human body. Contemporary materialists or physicalists in the philosophy of mind, at least in favouring or supporting its analytical idiom, largely equate the mind with the brain. To be sure, there are persistent disagreements as to whether the identity is a strong one (type identity) or a milder one (token identity), involving a relationship of supervenience and subvenience, and if so which type (e.g., strong or weak supervenience), supervenience having enjoyed much more widespread discussion, the term definitely emerging as one of the great philosophical mantras of the day. But if the mind 'is' the brain, what happened to the body? Well, it seems that in this context the (human) body is just the vehicle of the (human) brain or brain-mind (the mind supervenient on the brain, or rather mental states supervenient on brain states), just as in another context the body is, in Dawkinsese, a widely shared though not unchallenged parlance, just the 'vehicle' of the genes, the conjunction of these two theses making up what one might call *the thesis of the double vehicularity* of the human body, or perhaps for short 'the V2 thesis' in philosophical anthropology.

The integrity of the body or the wholeness of the human being is not a prime concern of these schools of thought, if not actively denied or minimized; for them, vehicularity or modularity (also a major theme in the contemporary philosophy of mind though the question of its scope is particularly controversial, and a master metaphor of contemporary social thought drawn like so many others, from the computational template)[8] serve well enough. Yet a positive conception of the wholeness of the human being can be and indeed has been formulated in other than the substantialist terms of 'rational psychology', in functional idiom, as in Renouvier's conception of the 'total human function' (as part of his critique of faculty psychology was a forerunner of the modular view of mind, and the implied view of the person as a bundle of faculties). This was a wider thesis than the fashionable 'holism of the mental' in the philosophy of mind, more like a *holism of the human*. This is not currently fashionable in philosophy or the social or 'human' science, to put it mildly, though in this form it was deliberately non-substantialist, and in tune with an ontology of events, phenomena, relations and functions; (the move 'from substance to function' as the vector of critical rationality was made in Renouvier's philosophical teaching some decades before it became a slogan in German neo-Kantianism, and not only).[9]

The body is, for the time being, regrettably, the human brain's only known life-support system. I say 'regrettably', because since the 1920s there has been speculation among scientists on providing it with another vehicle, another sort of life-support system, dispensing with the human body as we know it, to enable the brain to flourish, or rather, to carry on its real work, without all this redundant flesh. For the more radical, the brain is still too much 'fleshware' or 'meatware' to be transposed *in silico*.[10] The viable, operative, brain-without-a-body, it must be stressed, is no mere recent science fiction (SF) fantasy. Indeed, the operative brain-without-a-body appears most often as 'the brain-in-a-vat' topos which appertains to both the analytical philosophy of mind and science speculation as well as to SF. The brain would still require vital nutrients – that is what the 'vat' is for – but not necessarily or no longer a full-fledged body. Note that the brain-without-a-body would also be, of course, given the mind-brain identity, in one form or another, the mind-without-a-body, even though cast in atheological terms. The soul, the rational soul, has long gone, of course, like spirit, and sometimes even consciousness, or 'the Cartesian theatre'. Curiously, no one speaks of 'the mind-in-a-vat', though; if the brain is in a vat, and the brain is the mind, where else, if anywhere, is the mind? This is a congenial outlook for those who see the cognitive operations of the brain as the ones of any cosmic significance. To put it another way, they are the only ones who are not purely parochial, and basically not interested in any others, except insofar as they are necessary for optimal cognition and the accelerated advance of cognition in the future, in the sense of the technoscientific cognition of reality. In other words, this entails command over and unceasing transformation of the natural world, on Earth and as soon as possible beyond, without limits or controls.

Yet, once upon a time – indeed, not so long ago – the mind *was* the body.[11] Or rather, the mind was behaviour and nothing else, in one of the great watchwords of the behaviourist movement. Yet behaviour (overt, public) was essentially the behaviour of the body, or rather the acting body, the body in motion, the gestural, postural, kinesic, locomotional body (in research experimentally induced).[12] In philosophy the 'body epistemic' is often construed in a rather narrow fashion; for Quine, as he put it so many times, the *(epistemic) ur-phenomenon*, so to speak, of his naturalized that is, psychologized, behaviouristic epistemology is the sensory 'surface irritation' accruing to or the stimulation impinging on the presumably ex ante passive body's sense organs, presumably the outer senses rather than kinesthesia, all the way 'from the amoeba to Einstein' (Popper 1999, 73).[13] Whitehead, in his strictures on Hume's account of sense-experience, drew attention to the massive diffuse background of proprioceptive sensation as something that Hume had completely failed to take into account, truncating his vaunted empiricism thereby (Whitehead 1928).[14] In the pragmatist tradition Quine would generally be affiliated to, there would be a bias towards regarding the 'body epistemic' as the active body, in the form of a grasping or handling or manipulation of things, the haptic or haptic-optic mode of engagement in the world, rather than the more contemplative, exclusively ocular, mode, bearing in mind the Rortyan strictures on classical epistemology and its dream of grounding the knowledge enterprise of mirroring the world in inexpugnable foundations (*fundamenta inconcussa*), if not epistemology *tout court*.[15] These strictures follow on the lines of Dewey's aspersions on the 'spectator theory of knowledge', the contrast being between the 'spectator' and the 'actor', though this dichotomy seems a little out of place in contemporary arts. Dewey perhaps had in mind if not the *theoria* of pure contemplation of Aristotle (in the conventional understanding), then somebody like a very well-behaved spectator of classical chamber music or Wagnerian operas rather than the participative, agitated, hyperactive spectators of many kinds of rock events, for instance, or of experimental theatre and contemporary arts that invite role-sharing by, or role-reversal between, artist and spectator, or even inviting the destruction of the work by the active spectator, for that matter. 'Spectating' in the old sense, which may perhaps have only come about in an institutionalized fashion from the middle of the nineteenth century in the West as a new 'technique of the body' or a complex of such techniques and aesthetic stance, is now perhaps a lost art of the arts.[16]

Generally, in epistemology there has been a tacit or explicit axiology or value-hierarchy of the sensorium, privileging some senses or sense-modalities, classically sight, whose praise has been sung by many though by no means all philosophers since Aristotle, and the distance senses as a whole and downgrading the others (Jonas 1966).[17] The 'body epistemic' as a sensing body, is always markedly inegalitarian even monohierarchical for the epistemologist, it seems, though no different essentially in its inegalitarianism from the perceptions of the 'body technological' from the engineer's or the designer's point of view. Lately, in considering the problems of manned space flight engineers have been reviewing the serviceable possibilities of *all* human senses, even of somesthesis, through a kind of technological pressure towards sensorial parity of consideration.[18] Just as there has been a *scala naturae,* a well-ordered chain of beings, and a conventional hierarchy of the fine arts, there has been a rank-ordering of the senses in epistemic as well as in other terms. Although scattered, and the object of a variety of disciplines or strands of scholarship, there has in fact been something like a tradition of what might be called *sense-criticism,* reflections and notations at least, if not often very thorough or systematic accounts, on the roles and ranks of the various discriminated senses in the economy of cognition, the economy of feeling and the over-all scheme of human life. Although much of this sense-criticism falls largely within the mode of epideictic rhetoric, as in the nobility of sight *topos*, implying a determinate hierarchy, there has also been concern for the negativities of the senses or of some senses and even challenges to the nobilitation of the visual sense and the exorbitation of its role in epistemogenesis. An area of investigation in what we may call *comparative sense-criticism* would be to determine how far various eulogistic or dyslogistic themes in overt sense appraisal or implicit value judgments and standard tropes bearing on these themes are more or less salient in different national traditions in different historical periods, in philosophy, literary theory, art theory, psychology, or even theology.

But just as there have been challenges to these scales, there have been historic rebellions too against this hierarchy, and indeed social radicals have challenged the dominant sensorial hierarchy as part of their attack on dominant values and institutions. In academic work, one finds a comprehensive cognitive, ethical and aesthetic critique of visualism, of the ocular-centric stance, in the French tradition. This is demonstrated quite remarkably in the thought of Bachelard (no enemy of science!), and later in Sartre's famous depiction of the reifications of the gaze (the 'looking at the waiter' case), in Ellul's critique of technology and its constitutive outlook, in Lacan, in Foucault's account of the 'medical gaze' (*le regard médical*) and in his later works, most famously in his invocation of Bentham's panopticon (which reintroduced the word and its variants to the world lexicon of the *bien pensants* in the human sciences),[19] and French feminist excoriation of 'the male gaze'. This tradition of relative negative appraisal of visualism or specularity ranges widely, as can be seen from the range of names cited, and goes back even into the eighteenth century.[20] The obsession with visualism and anti-visualism was for a couple of decades fostered by Sartre's influence, though this specific strain had little impact on the social studies, but its main impact today, at least in sociology, stems mostly from Foucault's influence in sociology. This can be witnessed in the formulation of various concepts of reductive, decontextualizing, appropriative, objectifying, reifying, degrading, *de haut en bas* socio-typical and typifying 'gazes',[21] such as the 'tourist gaze', for instance, though the 'sociologist's gaze' – or indeed 'touch' – appears to be exempt, or at any rate never invoked. It is not always clear whether the rehabilitation of the life-world of oral–aural non-literates is in question. This McLuhanite theme has not had much impact in France – though at times not just sight but other senses partake of the same kind of negativity, as in Sartre's obsession with *le visqueux*.[22] But there seems to be a general critical movement of repulsion towards analytical rationality and technocratic attitudes through these varied works, though almost everything that can be disvalued in generic terms can, so to speak, be brought under the lenses of ocularphobia: patriarchy, phallocentrism, occidental reason, bourgeois order, power at large, modernity, scientific reductionism and hubris and the process of civilization.[23] The more sophisticated discriminate in Heideggerian fashion between different kinds of seeing: the wrong sort (the 'gazes' of jailers, males, tourists, Westerners), and the good sort ('alethic' seeing), non-reductive, patient, inclusionary, humble, within the horizon and non-Icarian. As far as I know, all these critics of visualism were sighted, though some lost their sight towards the end of their lives.[24]

The association of the visual sense with the perversions of science need not stem from Heidegger or 'Heidegger's children' (Wolin 2001), as it would have been clear in the cases of Bergson or Bachelard, wholly untainted in this respect, and can be found in quite different cultural contexts. Take the following quote from Kenneth Burke, a rather singular American scholar: 'the entire scientific attitude, which has flourished concomitantly with the growth of a predatory society, tends to think of understanding as *ocular*, a matter of *seeing*'. It is true than in recent years comparisons have been made between Burke and Heidegger, though they wrote in complete independence. Nevertheless, for nearly a hundred years there has been talk of a 'loss of visualizability' or 'picturability' in theoretical physics, though recurrently also it has been disputed whether this loss is as deep and irreparable as it has been made out. Again, from the late nineteenth century, psychologists claimed to have discovered the existence and importance of 'imageless thought' not least in scientists, and of course the images they had in mind were primarily though not exclusively, visual.[25] Historically, too, auditory and even musical or musical-instrumental models have been important in the history of physics (even outside acoustics) and cosmological theorization (think only of the *topos* of 'the music of the spheres') even if perhaps they should be cultivated more, as a corrective to the strong visual bias in our civilization (though *theoretical* physicists at any rate have often been musically cultivated).[26] It is curious that Shannon in his original constitution of information theory, contradistinguished 'information' and 'noise'; thus an auditory metaphor presided at the baptism of one of the most important categories of thought today, even if as a negative correlate (understandably in the case of someone working for the Bell Telephone Company).[27]

To return to behaviourism: no primary reference or special causal attribution was made in this perspective to micro-structures of the body (except in rather gross fashion), those of the brain, in particular, not playing a key role and not ascribed a paramount explanatory status in this scheme of thought. This paramount focus on the overt, outer body warrants the label given to this outlook of 'molar behaviourism' by the philosopher C. D. Broad, in contradistinction to 'molecular behaviourism' of intra-organismic, endosomatic research programmes of neuroscience and psychophysiology (though psychophysiological physicalism or psychophysiological reductionism nowadays would be less confusing terms, since the advent of molecular genetics and biology and the powerful reductionist research programmes arising from them; Broad's contrast was methodological rather than substantive, and 'molecular' was not taken very strictly in physicochemical terms).[28] Molar behaviourism 'black-boxed' the endosomatic or intracranial structures, processes and mechanisms of the behaving body, hard as it may be to believe today when another style of reductionism, that of micro-reduction, and the privileging of the molecular, rather than the molar, level of explanation in the life sciences are all the rage.

Now the basic intent of molar behaviourism was to bracket the existence of consciousness, or of a nonphysical mind, if not to regard the very question as barred *ab initio*[29] in any would-be science[30] and to secure testable hypotheses for all kinds of human behaviour, public happenings all, normal and pathological, without invoking introspective or introspectible data as either sources of hypotheses or test statements for psychological hypotheses (which could only be confirmed or disconfirmed by behavioural observation statements). In principle, the totality of animal and human behaviour, the behaviour of all putative psychological beings, would be encompassed within the same basic scheme of explanation, and within a uniform observational language, which would be devoid of intentional, purposive, teleological terms: 'behaviourese', a language of molar behaviour descriptors. In fact, one might say it is a cognitive 'body language' or 'language of bodies', a language in which only overt, outward bodily happenings and performances that make up, as it were, the public – in the epistemic sense – 'stream of behaviour' in daily life are taken into consideration. Leslie White, the founder of the cultural materialist tradition in American anthropology, often suggested replacing portentous nouns by verb forms in order to reduce their symbolic charge and make them more amenable to rational scientific discussion.[31] This was in order to discourage their proneness to substantivalism or substance-talk, roughly what came to be called essentialism, the alleged illusion of permanence and depth, or underlying, hidden, rock-bottom properties that can be defined once and for all. Thus, he suggested replacing 'mind' wherever it occurs by 'minding', 'science' by 'sciencing', and so on – though in opposite fashion that staunch nominalist, Bentham, would replace verbs by nouns wherever possible.[32] In a similar vein, one might suggest that 'body' (in the sense of human body as well as in other senses) could well be replaced fruitfully, often, if not always, by 'bodying', or 'bodying forth' in a slightly archaic but still recognizable idiom, or simply 'behaving' or 'behaviour' (molar or molecular). 'Behaviourese' de-substantivizes the body (and the mind or the mental) and verbalizes or 'adverbializes' it by construing it only as the site of acts or episodes of behaviour, as occurrences or occasions of 'bodying'. We should mention, though we lack the space to discuss in detail, the dispositional properties version of philosophical, conceptual–linguistic, adverbially addicted behaviourism, advanced with such brio and verve by Gilbert Ryle in his *The Concept of Mind*, published in 1949 (which inaugurated a long-running fashion for language-philosophy or conceptual analysis books – none of which matched its impact – entitled 'The concept of . . . .', but not, as far as I can recall, 'The concept of body'). The contrast between dispositions (to behave in such and such a fashion) and occurrences (or episodes), although not original with Ryle, plays a crucial role in exorcising the allegedly Cartesian or Cartesian-derived 'ghost-in-the-machine' image of the mind, that is, removing *both* the complementary 'myths' of the ghost *and* the machine, both of civilizational and not merely academic, importance, for, according to Ryle, these intertwined myths had been part and parcel of the dominant intellectual outlook in the West since the seventeenth century.[33] To overthrow the 'ghost in the machine' (implicating also, concomitantly, 'the bogey of mechanism') was therefore something of an intellectual revolution with broad implications for the conduct of our political life.[34] Cartesian-substance dualism may have been overthrown, but nothing like Ryle's integral molar behaviourism, where mental and bodily functions are equally glossed as dispositions to behave in such-and-such a fashion, has replaced it as central in our dominant intellectual outlook. It is not clear whether any conception of the mind informs our domain intellectual outlook, though it is perhaps fair to say that it is becoming more and more physicalized, nor how far this may affect the climate within which public affairs are carried on, since we still affirm what Durkheim called 'the cult of the (human) person'.

In principle, this 'behaviourese', this language of bodily performances, this language of bodying or bodying forth, could be deployed in any field of scientific inquiry the behaviourists saw as essentially restricted to observables. Behavioural inquiry could be addressed to all (molar) bodies, aiming at a science of universal bodyhood or universal bodying (as in 'adverbial materialism'), against the background of a pansomatist metaphysic (more or less mitigated).[35] In this light, one could well imagine a behavioural physics or a behavioural biology (ethology has been close to this hypothetical model of behavioural biology), except that the sciences of physics and biology, with an abundance of what are now called 'scopic' instruments, mathematical formalisms, experimental procedures, and a long and successful history already existed. Consequently, given this preemption, only restricted cognate fields such as behavioural optics and other kindred behavioural studies more or less in behaviourese, connected with all the human senses, of obvious practical relevance were ever developed in the interstices of the human-physical studies. The expansion of the molar behaviour research programme, whether we call it 'metaphysical' à la Popper, or 'scientific' à la Lakatos, beyond psychology (human or animal), had rather to look to undeveloped domains, such as the social sciences.[36]

In cultural anthropology, Marvin Harris, the leading contemporary exponent of cultural materialism, outlined the kinds of protocols a molar observation language should follow, eschewing all reference to goals and intentions, so that locutions such as 'in order to', 'for the sake of', 'with a view to', and the like, would be systematically deleted. An example is the proposal by the anti-clerical, historical materialist Neurath for an 'Index Verborum Prohibitorum' of metaphysical terms, locutions and questions, a would-be manual of Epistemological Correctness. Harris's book, *The Nature of Cultural Things*

(1964), is little-known, but not unrepresentative of a large body of thought. Despite a promising start, a (molar) behavioural sociology research programme along similar lines never quite made it to the mainstream. It is hardly ever introduced or even mentioned to students in most universities in the Western world, in part because of its naive empiricist or operationalist epistemological prepossessions ran against the grain of theoreticism in sociology, which survived and indeed developed even further after the demise of functionalism but, more importantly, owing to the strong hermeneutic commitment that has come to prevail (under the influence of language philosophy, Saussurean general linguistics, semiotics and hermeneutics proper). In economics, if anything, the rejection not only of behaviourism as a doctrine or scientific research programme in psychology but of the 'molar behaviour stance' in general went even further, indeed, much further than in sociology, and a research programme comparable in scope and productivity to that of behavioural sociology never developed as far. A partial exception must be made for classical American institutional economics, the avowed behaviourism of which – or rather behavioural stance, since they did not subscribe to a specific doctrine of behaviourist psychology, despite their general sympathy with the behaviourist movement – was of a social rather than a physicalist sort and, although rejecting teleology, spirits and Cartesian minds, did allow for mind as a social category, as symbolic interaction, and in any case it failed to alter the course of economic thought.

But, even in economics, Vilfredo Pareto, in his concern for value-free science, at times seemed inclined to a kind of behaviourese, as when he replaced the concept, or at least the term of utility, as too subjective or mentalistic, by the term 'ophelimity' (largely forgotten), something close to 'revealed preference' and cardinal or ordinal behavioural utility. Lately, 'experimental economics' (economics has been the most consistently non-experimental of self-certified empirical sciences), which is the nearest thing to a 'behavioural economics' that has as yet come to pass, has concentrated on such things as the choice-behaviour of rats, though speaking unabashedly of their 'utility functions'; not surprisingly, according to experimental economics, rats behave as if they were cognizant of rational choice theory (as pigeons had already complied with operant conditioning theory for Skinner, and sometimes also for economics). Parenthetically, one might ask if there is virtually no behavioural economics, why is there not an 'economics of the body' comparable in volume of production to the sub-disciplines of the anthropology or the sociology of the body? After all, there is or there could be an economics of everything – an economics of the performing arts, an economics of sport, an economics of health, of the environment, of prayer, of knowledge and so forth, and there already are books on the 'economics of the *mind*' on Hayekian lines. What we might call the *economorphic* mode of thought involving the closely interrelated concept of opportunity cost, the marginal principle, the emphasis on substitutability in production and consumption (so crucial that we might characterize the basic concern of economic theory as not so much with a Pure Logic of Choice, as Hayek put it, but as a General Logic of Substitutability), or in more elaborate forms, such as rational choice theory and cost-benefit analysis, exercises extraordinarily influential, perhaps even stiflingly so. It is the general character, the general logical form, of such accounts that is most pervasive; they need not be couched in substantively economic terms or stress the causal priority of economic variables. Even the concept of externalities has now begotten by analogy that of 'internalities', referring in the case of the economics of the mind to the secondary effects within the person or mind of one's choices, relevant no doubt to an economics of the mind, though presumably this could be extended further to the organic or psychosomatic side-effects in oneself of one's actions in an economics of the body.

Perhaps, after all, there has been a methodological bias against the body as a suitable area of investigation in the conceptual framework and basic analytical strategy of mainstream economics (and indeed amongst the bulk of the very numerous dissidents within or outside the profession, though possibly there may be exceptions), with its rather ethereal rational economic agents, but a sub-discipline of the economics of the body is sure to appear formally any day.[37] This might well put the Beckerian economics of the family – really of the household – in its place, though quite different modes of the economics of the family could be worked out on a variety of substantivist grounds in lieu of the two-atom model not so much *égoisme à deux*[38] as the ever-revocable conjunction of *deux égoismes*, which is all basically all that Becker has to offer. And it is not too difficult to envisage some of the paths the economics of the body might take.

Wherever there are markets, formal or informal, there economic analysis has a role, and presumably by the same token '*body economics*', which might address the markets or quasi-markets in the most tradeable and traded bits and pieces and services of the 'body economic', which comprise mostly body parts, sperm (what has been called 'techno-sperm' but could also be called 'market sperm'), eggs (similar considerations would apply here: 'techno-ova' and 'market-ova'), womb leases, embryos and genes, and last but not least blood, though this has escaped commodification in some countries. Classical contractarianism anticipated these developments to some extent in its way of framing the ethical import of social relationships; one has only to recall Kant's characterization of marriage.[39] Its key sub-discipline might come to be called *geno-economics*, dealing with the market dynamics of transactions dealing with genes and their alterations and transpositions (presupposing therefore the expanding capacities of *geno-engineering* and of bioengineering in general). We have been witnessing after all the emergence of the 'universal body shop' and 'the genetic supermarket', local and transnational (the latter a concept already advanced by the market libertarian political philosopher Robert Nozick in his treatise *Anarchy, State, and Utopia* of 1974).[40]

Indeed recent developments have drawn attention to a passage in Rawls's *A Theory of Justice* (1971), previously little commented on, it seems (though it is hard to believe, given the immense amount of Rawls's commentary over several decades) where he suggests that one may have an obligation to improve one's natural assets, against the background of technologies that may improve the genes of our progeny so that they can start off not only with the best environmental advantages we can give them but also with the best 'natural capital' (genetic or natural human capital) in terms of body and mind that money can buy: strong bodies, beautiful bodies, ever so tall, and every other desiderated quality.[41]

Thus, we would have a market in future bodies or future genotypes – often taken as synecdoches of whole bodies – or at least in their potentiation or genetic enhancement through superalleles: a topic for the geno-economics of future persons. There will be perhaps geno-economic, natural human capital experts, consultancies, advice columns, New Geno-Economy firms, Government departments with programmes of natural human capital or genetic capital investment, and surges of natural capital accumulation in the affluent populace.[42] What one might call *genic (or transgenic) perfectionism*, the drastic improvement of our condition via genetic engineering or positive technological eugenics (technologically improved genomes) has become in effect perhaps the most salient and appealing form of perfectionism in the long and variegated history of this tradition in Western imagination and utopianism, though now focused almost exclusively on individuals rather than populations, enabled through the market rather than through the state, without any inclusive social goals in view, micro-utopias of micro-eugenics.[43]

For a while in the fifties and sixties of the last century the social sciences were indeed often called in the United States above all 'the behavioural sciences', but this was not exclusively, or even predominantly, because of the impact of behaviourism on psychology, the behavioural research programmes in a variety of fields, or of the persuasiveness of the molar behavioural stance as a scientific outlook. Rather, this was a way of neutralizing the widespread suspicions arising out of the fear of communism, socialism, socialized medicine, socialization of the means of production and the like, all very much 'social' things, which turned the expression 'social sciences' into what would nowadays be called a politically incorrect one, at least for a decade or so; other locutions such as 'policy sciences' or 'sciences of action' did not prevail.[44]

With the demise (partial, to be sure, but effective enough) of molar behaviourism as doctrine, methodology and metaphysic, and the waning of the molar behaviour stance, the body appeared to have gone with it, too, at least as a central category of thought in the human sciences. It took three stages of scientization of the human, at first in methodological guise and later with more overt ontological implications. First, there is *disensoulment* under the slogan 'psychology without a soul', although initially this programme purported to be solely a methodological, not a substantive or ontological stance.[45] Secondly, what we might call *dismentation* or rather another try at this after the demise of the 'psychology without consciousness' (molar behaviourism), that is, the dissolution of mind and consciousness as nothing more than epiphenomena at best, if not simply deleted as illusory according to 'eliminative materialism', currently a vigorous, vocal and well-funded programme in neurophilosophy or neurocomputational reductionism.[46] Finally, nothing was left to accomplish in biopsychological studies but *disembodiment* of one sort or another in order to get even more scientific, more mechanistic: not, it is true through corporeal eliminativism. Talk about bodies is not illusory, as it would be for, say, Leibnitz, but through the claim that mentation does not require an organic body but can be carried on with other types of material support (the thesis of materialist functionalism in the philosophy of mind) and in a technological perspective, whereby the natural, organic body becomes an object of re-tooling, re-engineering, mechanization and replacement by another, better designed, superior, electromechanical entity or super-entity. The body has, so to speak, been under strenuous conceptual attack in contemporary currents of thought.

Yet as late as the sixties of the last century, the body had been brought back into left utopianism by, among others, the scatological-mystical Norman O. Brown (for whom the essence of the human being is desire), and even into Marxism in the Freudo-Marxism of H. Marcuse, a synthesis of the young Marx of the Paris manuscripts and the old metapsychological Freud, with recovered dashes of Fourier, in which a biological, biopsychological basis for socialism, a 'natural' standard for the ethical critique of generic culture or civilization, had been rediscovered. As it turned out, it was but a Californian moment, which lapsed into what Marcuse had already warned against, 'repressive desublimation', and no invocation of 'plastic sexuality' will change matters. In their concentration on sexuality, the bringing about of something like Fourier's 'New Amorous World', though no longer in a Newtonian cosmos, they rather overlooked, for example, Fourier's admirable concern with 'gastrosophy' (though the fashion of biofeedback may not be unrelated).[47] Contemporary technological utopianism, by contrast, even in its left-descended versions, is very much against the body, and its major scenarios involve its supersession, though in genetic perfectionism body improvement, in its twofold vehicularity, is still on the interim agenda, prior to the post-biological era.

There are various ways in which the elimination, the extrusion, the virtual disappearance or at least the sharp downgrading of the body, as a fact or a value, as a prospect or a tendency, has been envisaged in very sophisticated ways in contemporary cultural studies and technoscientific research programmes, in technoscientism and technological utopianism. It is not at all far-fetched to claim that a kind of systematic somatophobia (the aversion to the natural body), or perhaps a curious sensory deprivation syndrome especially regarding somesthesis, a philosophical, possibly purely philosophical, body-loathing – though there need not be anything personal in it, so to speak – appears rampant in the academy, media labs, the technological avant-garde and a number of technology-orientated social movements, some ostensibly radical, though by no means without partial countercurrents.

The contemporary world is full of materialists who rejoice that matter has finally been abolished and that, instead of matter, information is everything and everything is information.[48] They include cornucopian economists who claim that wealth is immaterial, professed sensualists who loathe monkish or Calvinist asceticism and yet want to overcome the grossness of the (organic) body once and for all, practising and theoretical hedonists, egoistic or not, who desperately want to be disembodied and their brains 'uploaded', irreligionists who claim to have discovered the secret of non-organic omniscience, and atheologians who believe in the Omega Point as a matter of rigorous science (and have got hundreds of pages of equations to prove it). They also include this-worldly unbelievers who cannot wait to leave the blue planet, secularist *bon viveurs* who are, metaphysically speaking, sick and tired of their corporeality and want something better, much better, staunch believers in the selfish gene whose genes unselfishly appear to command them to get out of the body and even out of the reach, or would-be 'off the leash', as E. O. Wilson likes to put it, as he staunchly believes you cannot ever leave the body, no matter your individual or collective delusions of causal agency or self-transcendence of genes, and into non-genetic computational immortality. To be sure, they do not necessarily sense any cognitive dissonance in their positions, but at least some readers may feel that there is something paradoxical in their outlooks.

A very succinct survey of the modes of systematic, theoretical rather than religiously enjoined or ascetic, somatophobia and of the concomitant technological projects of discarnation that have emerged in recent decades would include the following:

1. The claim that we 'lose the body' and become so to speak 'discarnate' beings, unconfined by a known definite spatial location once we enter cyberspace. This had been already advanced by the anthropologist Edmund Carpenter (1974) on a broadly McLuhanite perspective decades ago. The comparison in this context with angels who are also bodiless was more farfetched, as angels had other more important properties than bodilessness or placelessness. In any case we are still bodily, incarnate, in real space if not in virtual space, in real time, indeed we are always and necessarily at a particular place, and no two bodies of our kind can occupy the same place at the same time, nor can the same body occupy two places at the same time, unlike some other kinds of material entities, such as microphysical ones.[49] As F. Waismann, better known as a philosopher of mathematics and Wittgenstein groupie, pointed out, 'I am here' is always and necessarily true whenever and wherever uttered (a peculiar synthetic a priori truth as he put it), what we might call the '*lococentric* predicament'.[50]

   Our bodies have a most intimate connexion with ordinary physical spatiality, while our minds have a peculiar disrelation to it. It seems that some contemporary writers would practically equate spirituality with information technology virtuality, at least as its rational shell, to paraphrase Marx. But what about the angels' nonstop, extraordinarily beautiful music in praise of the Lord? Nevertheless, the analogy with angels was launched a few decades ago, and the reconsideration of the traditional doctrines of angelology have treatises that have long been gathering dust in obscure corners and dusty stacks of the great Western libraries. The more erudite cyber-theorists are able to display an impressive knowledge of Jewish, Christian and Muslim versions and in cyber-doctrines today; we have not only a systematic cyber-theology, but also a cyber-angelology in process.[51] This cyber-angelology maps out the world beyond the body, the world after discarnation, and the higher forms of post-human being after disembodiment of one sort or another.

2. There is no doubt that 'informatics exalts the creative act, and especially the form of creation that appeals to the mind and not to the body. Creation without body is yet another value of informatics' (Breton 1990, 52). There seems to be in any case an 'elective affinity' between informatics work and the typical contemporary project(s) of discarnation entertained by some members of the technological avant-garde.

3. The doctrines of social constructivism reduce or unmask all forms of knowledge, all kinds of knowledge-claims, scientific or other, to or as versions without end of the ever-renewed, all-encompassing social construction of reality. There is no fact of the matter that would make a statement – any statement – true or false, except within a form of social life, and just for that form of social life. Many years ago, the philosopher R. B. Perry, an epistemological realist, had formulated 'the egocentric predicament', meaning that no matter what we think, however great our efforts at attaining or eliciting inter-subjective invariance, our beliefs still belong to the

knowing subject, the cognizing ego. Contrary to what is sometimes stated, Russell did not coin this expression, though he did use for many years the formula 'egocentric particulars' for what are more often called deictics, indexicals, demonstratives, shifters, token-reflexives, and so forth; for him 'this' became the basic irreducible egocentric particular, rather than say, 'here', or 'now', 'you', or 'I' – which would be egocentric in double fashion (Russell 1940). In social constructivism we have, as it were, a socialized or comparticipative version of this perplexity, although still fraught with the confinements of indexicality. Indeed, this approach is critically indebted to the ethnomethodological emphasis on the indexicality of all utterances, what we might call 'the sociocentric predicament'. This is the self-enclosure trap that occurs where the knowing subject has been replaced by the cognizing consociates (usually mixing in a lab, so far from Nature, so near sociologists). Ludwik Fleck's expression 'thought collective' is widespread, though its exponents do not call it a predicament, and perhaps would not even consider it to be such, a thorny philosophical problem, or even a problem *tout court*, but rather a solution.

What concerns us here more specifically is the 'over-socialized', the radical social constructivist, view of the body, the view that the body is 'nothing but' social constructs. It should be noted that there are no pan-human universals of social construction, only the universal truth that all is socially constructed and could not be otherwise, adopted by some, but only some, feminists (often combined with Foucauldian biopolitics).[52] It has been reported more than once by feminist theorists of a different persuasion that to advance, however timidly, the view that it may be appropriate sometimes to consider organic bodies as such, even sexual bodies, not simply and exclusively social constructions of them, at a seminar, can arouse not only the normal academic principled disagreement, or polite unanimous rejection, but expressions of blazing hatred, or at least sustained booing. Indeed, the watchword for this school of thought might be 'no bodies anywhere!', just the social construction and the politics of putative 'bodies'. Here politics is nothing but power struggles, even if 'power' begets 'knowledge' of, or rather, knowledge claims about, 'the body'. The existentialists had already taught that the *pour soi* was no-thing, now we learn from the social constructionists that we are no-body or no-bodies but similarly all *praxis* with no *hexis*!

4. In the late nineteenth century the biological theorist August Weismann, in his struggle to overthrow all vestiges of, or at least to deliver a coup de grâce to, the doctrine of inheritance of acquired characteristics (which Darwin, unlike contemporary 'Darwinists', had not entirely ruled out), drew a sharp distinction between germplasm and soma, the factors of heredity, on the one hand, and the mere body, on the other, a dichotomy which, in its general form, has proved a lasting one For this biologist the germplasm was immortal, the soma mortal. Bodies come and go, but the germplasm remains, quite unaffected by the living bodies it has inhabited; the germplasm essentially and enduringly is the same for a 'natural kind', the soma individuated and unique. It is interesting to note that Durkheim noted favourably Weismann's then fairly recent approach in *Le suicide* (so much for the alleged French obsession with

Lamarckism). He drew an analogy between the 'social type' – collective, enduring, prior in some sense to individuals and generations – and the germplasm, on the one hand, and the soma, or what would be called now the phenotype, with the individual, both mortal, secondary, particular, on the other hand. The autonomy of the social type would correspond to the autonomy of the germplasm. It should be pointed out, however, that Durkheim was no radical social constructivist as regards the body, for he viewed the human body as the principle of individuation, in the manner of the *principium individuationis*[53] of the Scholastics.[54]

Before long, the words 'genes' and 'genome' were coined, and now the standard distinction lies between genotype and phenotype. In the Dawkins formulation, to which we have already adverted, the genes, the carriers of information, are sharply conceptually contradistinguished from the phenotype, the body, as the 'vehicle' of the enduring genome. This merely vehicular characterization of the body disturbed even some biologists and philosophers of biological science. The philosopher of science, David Hull, suggested a change of terminology regarding a general scheme of 'evolutionary epistemology' (really a kind of generalized analytical Darwinism, applied to many different kinds of entities, natural and cultural, and not just organisms or other biological entities). According to this proposal, which has enjoyed some receptivity, for genes or other information-carriers such as 'memes', we should read 'replicators', and for the rather disparaging and passive-sounding 'vehicles' we should read 'interactors', for the Darwinian approach was and is essentially populational, independent of the ongoing controversy over the privileged or at least the relative roles and weights of the putative 'units' or 'levels' of '[natural] selection' (gene, phenotype, population, species, deme, etc.). In bio-evolution, this is not likely to be resolved soon – as witnessed by the inconclusive discussions of the recent Sober-Wilson approach to 'group selection' which, at any rate, has the merit of putting forth a strongly argued case for a new version of a discredited, indeed anathematized, idea.[55] At least on this view the living body, not least the human body, has something to do besides being a mere vehicle, a mere bearer and carrier of something far more important than itself, a kind of *Kulturträger*,[56] with the genotype standing in for *Kultur*, until such a time as it is replaced by another body-vehicle, and so on. It is assigned in this scheme the fun job (not to mention the vital necessity) of interacting with the wider world and other bodies/minds, increasingly of course with and through artefacts, within an increasingly anthropogenic, technogenic and virtualized world. We lack the space to discuss the still incipient Dawkins-inspired research programme of 'memetics', but: Do memes require bodies, just brains, or the functional equivalent of brains? According to the functionalist philosophy, mind brains are just one of indefinitely many potential supports for the mental, in which case would they not even require genes? What 'vehicles', carriers or supports do memes or, more precisely, meme-tokens, require in the most general terms?

5. Writing in a fervent speculative vein long before the advent of the contemporary molecular-informational approach to genetics and biology and indeed the computer revolution in science and technology, which he came to witness and participate in, the

physical chemist J. D. Bernal anticipated much current technoscientific speculation on overcoming the body as early as 1929. He envisaged a situation in the not-too-distant future wherein the growth of (scientific) knowledge could be impeded by our corporeal limitations. No matter: we could and indeed should gradually replace our sense-organs and our motor organs, our sensorium and our motorium, by advanced electromechanical instruments, sensors and effectors, built into our bodies, which would thereby become increasingly electromechanical rather than organic, a 'body technological' as well so to speak a 'body-for-technology' or 'body-for-science' – a vastly improved 'body epistemic' with vastly enhanced sensors and effectors, all for the advancement of technoscience.[57] By these means, the brain could access far more and better data, through vastly enhanced ranges of reception and manipulation of the outer world made possible by our sophisticated endosomatic prostheses (sensors and robotic apparatus). However, having largely disposed of the body, at least the natural, organic body, now but an electromechanical sensorial-motorial apparatus, there would be a problem still about the brain. On the one hand, it would need vital nutrients, hence the 'brain-in-a-vat' conception. On the other hand, since science and technology are essentially collective endeavours and Bernal was a collectivist in political thought (he died in the Communist faith) and a staunch believer in the enhancement of scientific cooperation and communication (a topic to which he devoted considerable attention), the brain-in-a-vat could not and should not remain a solitary knower, a sort of Crusoe brain, or Crusoe mind, in the manner of 'methodological solipsism'. Not being able to specify exactly how it could come about, Bernal still envisioned the increasingly intimate interactions of the many brains-in-a-vat in an ever more-perfect union of knowers, a *unio cognitiva*, a union in knowledge and for knowledge (technoscientific knowledge only of course), through which the Many become increasingly One, a collective epistemic super-subject, perhaps even a common sensorium, the common sensorium that was lacking even in the most organicist and holistic theories of society or doctrines of 'group minds'. One would be tempted to speak of 'symbiotization', but where is the *bios* in this field? It must be acknowledged that for Bernal the bonds between the brains-in-a-vat and their successors were affectual too, being bound together. He wrote in a particularly purple patch, with a love of which our own experience of sexual and mystical unions, of body fusion and states of pure consciousness such as mystics claim, can only afford the barest glimpses. Another way in which Bernal anticipated contemporary technoscientific speculation was in his insistence that the pursuit of knowledge, the pursuit of technoscience, demands that 'we' leave the earth forever and explore through ever more-sophisticated agencies in outer space and on and on.[58] At the very limit of the technoscientific quest, 'we' (i.e., the remote successors of the 'brains-in-a-vat' and our even more remote descendants) will become pure light: epiphany becomes, as it were, photophany.

Of course, all this only follows if we postulate that the vocation of the human being, the destiny of homo sapiens, is the maximization of technoscientific knowledge, its pursuit without end above all else, ever faster, and that everything that is given to us on earth must subserve this paramount goal. A goal that deserves a name: The Mandate of Epistemophilia or, better, *Epistemolatria.* The knowledge in question is, of course, technoscientific knowledge. Most emphatically, they do not recognize any other kind. This a widely shared and even deeply internalized presumption today, and hardly recognized as a presumption. More taken for granted, as if it could not be otherwise for anyone other than the most obscurantist (against science!). But is it really and truly rationally incumbent, even binding upon us as a species, to trans-speciate ourselves in this fashion, regardless of the price, regardless of all other values besides the maximization of technoscientific cognition and of its rate of advance? Yet the Imperative of Human Trans-Speciation with a view to mechanization, as a corollary of the Mandate of Epistemophilia or Epistemolatria, has been proclaimed by a number of significant voices within the scientific world, outside the medium of SF, from at least 1929.

It should be added that human trans-speciation, *still within the biological world*, through genetic or genomic engineering, and not as a leap into 'uploaded' mechanization, is currently contemplated by biologists and scientific visionaries. There are biologists less possessed by somatophobia or *ontological misanthropy* (human species aversion on the grounds of its radical, incurable inadequacy as epistemic or technological beings, since they are organically embodied), than the technological speculators we have been concerned with. They argue that we are still a young species against the background of natural history, a fragile species; thus, one might say, as a corollary, that we are too young and too untried to be fed so soon to these projects of technological trans-speciation (Thomas 1992). In any case, perhaps we should stipulate eliciting the 'informed consent' of the species as an absolute requirement prior to trans-speciation, certainly in the case of our species. For our species qua biospecies has become increasingly a laboratory for itself, under self-scrutiny, an experimental laboratory, a clinical laboratory, a factory laboratory – now, of course, with the best possible intentions. To say this is to be controversial, much more so than it would have been, say, a decade ago. For although 'species' is indispensable to biology, despite the considerable conceptual and meta-theoretical divergences on the criteria for optimal delineation of the category, it has been much disputed of late whether species-membership in general or membership of the human species in particular is a matter of any serious or overriding moral consequence, and indeed whether being human is a 'natural kind', fragile or otherwise, though we acknowledge the existence of a human genome. Yet we worry about species diversity and accelerating species extinction and the category 'species' still speaks *to us*, but not it seems *for us*, as we become a 'species laboratory' under permanent biotechnological self-observation and self-modification; though biotechnologically and not merely spiritually, we are no longer just a species-in-itself but a species-for-itself. To be sure, only some individuals benefit, but the major kinds of interventions bear not just on specific individual characteristics but on generic traits and capacities that implicate potentially everyone (as in the case of 'regenerative medicine', mentioned earlier).

6. Given the accelerated, exponential growth of computational capabilities according to 'Moore's Law' (duplication every 18 months), we may be confident, it is widely claimed, that we are approaching in the coming decades the transition from 'auxiliary artificial intelligence' (the only kind yet available) to the long-desiderated autonomous version. That is to say, the threshold of human-level intelligence will be reached. It is always

assumed, it seems, that we will not be able to skip from subhuman to above-human intelligence without going through a stage of parity between intelligent machines and ourselves. Beyond that, further accessions of cognitive power are on their way to the Ultra-Intelligent Machine(s) or something more symbiotic in human-machine terms, or more corporate, as a machine of machines.[59] But which will play the same eschatological role of a Singularity in Earth history and perhaps even in the wider cosmic history? This will understand us perfectly, encompass all our knowledge, but we will utterly fail to comprehend it. It will be far beyond our cognitive capacities, and we will achieve thereby the self-transcendence so longed for in some religions, or at any rate accomplish our obsolescence in the daily, hourly process of creative destruction that is the market economy.[60] There have been modest triumphs, such as the Deep Blue programme, which beat a Grand Master at chess once (though 20 years after it had been predicted as imminent by key AI scientists), but it seems there is far, far more, and far better, to come. Moreover, not only are computers now endowed with parallel and distributed processing, but they are becoming superparallel,[61] and through an extraordinarily rich topology of interconnectivity, their information-processing powers are further enhanced, leading in due course to the emergence of the computational World Brain or planetary Collective Intelligence or technological noosphere, where in fact the noetic should be read as info(rmation), as an infosphere.

Nevertheless, the topic of main interest here is what happens to the body and its synecdoche, the brain. The brain-in-a-vat motif, the brain largely dispensing with the (rest of the) body, continues to interest commentators and philosophers of mind today. Curiously the impressive bulk of ongoing publications on the philosophy of the mind is not matched in the universe of analytical philosophy by a comparable volume of publications on the philosophy of the body, certainly not under that rubric, and what appears is within the aegis of the philosophy of mind and neurophilosophy (increasingly in the shape of neurocomputational philosophy), not as a sub-discipline. Perhaps, having seized the mind/brain, the contemporary analytics willingly left the lesser question, the body, to Continental hermeneutical or existential-phenomenological or at least to feminist philosophy, a matter for *philosophes*, not for real 'philosophers', as they sometimes put it, though there are signs that the situation may be changing.

Insofar as the brain is regarded exclusively in informational terms, as a storehouse and as a processor of information, its information content and processing capacities could, in principle, be made available in, or transferred without loss to, another carrier or vehicle of information and information-processing, not necessarily organic or carbon-based. What is needed, and is not yet available, is an adequate technology of brain scanning to retrieve and transfer the brain's information in its totality. When this brain-scanning technology comes on line, so to speak, say within the next twenty or thirty years, all will be finally ready, even without waiting for the opportunity of really placing our brains in a vat. We can have our brains uploaded, without pain, and perhaps without feeling a thing, into more than humanly intelligent machines, our bodies and brains now *both* technologically obsolesced, superfluous to the growth of technoscientific intelligence, and therefore quite worthless, if not downright pollution in the infosphere (and pollution has to be eliminated). The uploading in this sense will be the final technological human discarnation, the consummation of human disenfleshment:[62] the 'fleshware' or 'meatware' of our gross, decrepitude-liable, mortal bodies, markers of our animality, of our paleo-biological condition. Our condition is paleo-biological because there is hope for a neo-biological civilization of androids and the like (Kelly 1994), which will give way to potentially immortal, super-intelligent or ultra-intelligent software. And, they say encouragingly, you can have the mechanical 'body', the hardware, of the colour of your choice, on your exit from the cornucopian, still-human, all-too-human Republic of Choice in which we dwell in the West. How strange and disappointing that in such lofty speculations these visionary writers lapse into the language of the car salesman, but perhaps just in order to bring out the utter banality of it all in due course. But the decisive questions of real choice would have arisen much earlier: in fact, right now, if not yesterday.

Notes

The present text takes up themes already addressed in my paper 'Hegel, Texas: Issues in the Philosophy and Sociology of Technology', originally published, in English, in 1993. It is also available in a Portuguese translation in a book entitled *Hegel, Texas e outros ensaios de teoria social* [*Hegel, Texas and Other Essays in Social Theory*] (1996), and subsequently in the article 'Ciência, técnica e utopia' [Science, Technology and Utopia] (2000). My first foray into questions of the body addressed death, dying and the corpse in 'Tristes durées', which was published in *Death in Portugal* (1983).

1 Since writing this I have noticed a recent book by Arthur Danto entitled *The Body/Body Problem* (1999).

2 Left-wing thought (as Léo Moulin has argued), and perhaps perfectibilism or utopianism in general – as John Passmore has claimed in his *The Perfectibility of Man* (1970), a historical survey and critique of perfectibilism – has been prevalently if not exclusively, Pelagian in contradistinction to Augustinian, regarding the primal character, the innate goodness or at least non-malevolence of primal human nature. Not only do we still use these Christian theological terms even when discussing non-Christian thought, but we have no other widely accepted categories in terms of which to handle this problematic in a post-Christian world. Kant is not, however, so neatly placed, for although a strong believer in the possibility of progress (intellectual and moral) and in the human capacity for moral regeneration in both individual and in species terms, and therefore a Pelagian in these terms, his thesis of the radical evil of human nature as inextricably intertwined with the constitutive capacity of moral freedom sets him apart from standard Pelagianism. This stance has been taken into account by Kantian democrats and democratic socialists, or 'ethical socialists' in the French tradition of Renouvier (certainly obsessed with the problem of evil) and those marked by his teaching where this thesis of fallen freedom becomes a postulate of practical reason, and in the German neo-Kantianism of the Marburg school. One must distinguish what I have called 'Kantian democrats' from what B. Yack called 'Left Kantians', that is, radical totalistic, utopian ones, postulating the possibility and necessity of a 'total revolution' in the human condition; and what Gurvitch called the 'Left Fichteans', essentially the Krausists, who were reformists rather than total revolutionaries. It should be noted that a certain strain of radical thought has been tempted by the doctrine of *apotacatasis*, notably in recent times by W. Benjamin. On the perplexing, subtle Kantian teaching alluded to, see Gordon E. Michalson Jr.'s *Fallen Freedom: Kant on Radical Evil and Moral Regeneration* (1990). Kant's 'crooked timber' metaphor (borrowed from Luther) has been widely cited and used as the title of one of Isaiah Berlin's collections of essays (1997); the attribution

to Luther is from the L. note to the French translation of Kant's *Anthropologie* (2008). The Stoics also used to say that a stick must be either straight or crooked, a person is either just or unjust, not on-the-way-to improvement as an Aristotelian would put it, though they did not use the image apropos of generic human nature (Laertius 1961, 37).

3 *Plenum* is a Latin term that means 'full'.

4 *Discreta* is a Latin term that means 'separate' or 'distinguished', while *thema* is a Latin term that means 'theme' or 'subject'.

5 This is in the well-known sense of Gerald Holton, that is, recurrent master images in the history of scientific thought favouring or supporting rival, mutually exclusive, modes of theorizing. The alternation of the emphases on discreta and continua is one of those noted by this author in successive periods of the history of science. Neither has ever entirely displaced the other. In a way, the corpuscularian-atomist tradition and the Stoic tradition of continua and the plenum were classical embodiments of the respective thema and counter-thema.

6 On Kotarbinski and Polish analytical thought see the still very useful book by H. Skolimowski *Polish Analytical Philosophy* (1967).

7 In a very interesting restatement of what a Neo-Stoic might reasonably hold today, in ethics particularly, the discussion of bodies or the general account of the nature and properties of bodies (what might be called 'somatology') is rather brief (Becker 1998). Of course, the influence of the Stoics in European thought from the sixteenth to the end of the eighteenth centuries was immense, especially through various formulations of Natural Law. Perhaps their single most-enduring legacy was the concept of the 'invisible hand', an expression that has enjoyed immense fortune and that Adam Smith derived from Stoic teaching (Macfie 1967), as he did his view of life as a game, though the former has become prevalently associated with economics and the dramaturgical view of life with sociology. F. H. Knight was perhaps the subsequent economist most devoted to the game metaphor in a broad sense in his essays on economic life and the ethics of competition rather than in the specialized mathematical senses of games of strategy. The broad metaphysical Stoic stance – plenist, continuist, determinist – continues to have strong appeal even without reference to their texts. Einstein would be a case in point. David Bohm worked out a foundational, super-holistic microphysics that would be continuist and determinist, and conceptions of the holographic universe continue to have some appeal. Recently the Stoic term *ekpyrosis* has been revived by cosmologists who envision many Big Bangs and not just a primal one and, hence, many Big Crunches. Their ethics, too, can be reformulated in contemporary terms as by L. Becker, though no major tract of commendation such as the Cynics have received has appeared.

8 See for example E. Spolsky's *Gaps in Nature: Literary Interpretation and the Modular Mind* (1993).

9 This is also to be found in Simmel's *Philosophy of Money* (1900), and specifically in his account of the metaphysical implications of a full-fledged money economy. Mauss's notion of 'total social phenomena' or 'total human phenomena' might have been shaped by Durkheim, and his influence has kept this tradition alive as also the 'personalist' traditions in philosophy, psychology and psychiatry. Some social constructivists regard the individual as made up of groups. It is not clear whether they thereby subtend an image of the mind as a bundle of social modules.

10 *In silico* is a Latin phrase that means 'in silicon' and is used to refer to something that is done on a computer.

11 In the long tradition of Christian Aristotelianism, but particularly in Aquinas and in Neo-Thomism, today the mind, or rather, the 'rational soul', is the 'form' of the body. This hylemorphic conception, rejected by Descartes, may have problems of its own, but so did the Cartesian substance dualism, practically the central target (one might even say the chief object of target practice) of the philosophy of mind, certainly in the last fifty years or so, and it might well address many of the problems of other current conceptions even if metaphysically uncongenial today. Michael Polányi's variant – though his general outlook appears to resemble

that of emergent materialism – was encapsulated in the formula that the mind is the *meaning* of the body.

12 This is speaking in universal terms, though a would-be discipline of 'proxemics' was founded by the American cultural anthropologist E. T. Hall to address the great range of variations in typical postures and stances in interpersonal situations as well as, more obviously, gestures. See also note 14 on Mauss and 'techniques of the body'.

13 We say five *external* senses, because, particularly in the eighteenth century, an 'internal sense' was posited in addition, in different ways by a variety of thinkers. Indeed, at least half-serious metaphorical talk about an inner sense or inner organ of perception can be found in important thinkers from the eighteenth century to the present day. This is whether the organ of perception in question be one for moral verities (as claimed by moral intuitionists such as Hutcheson) or for mathematical truths (by mathematical Platonists, including Gödel). Also, recurrent reference has been made to a 'third eye' ('the pineal eye' in Bataille, the pineal gland being a vestigial eye) or to a 'third ear' in mystical writers; Descartes's 'the mind's eye' was not meant in any such way. The modular theory of mind posits in a sense the equivalent of sense-organs for language, music and mathematics, if not for perceiving in a quasi-visual sense, for processing different informational modes.

Contemporary psychophysiology discriminates many more than five 'senses', but still cleaves to the number five in terms of five perceptual systems (the basic orienting stem, the haptic system, the savour system, the auditory system and the visual system). According to J. J. Gibson's classic account (1966), the five perceptual systems are matched to fifteen senses; thus the savour system includes five senses on its own right, namely, smell, taste, tactual, thermal, pain; and the haptic system includes tactual, kinesthetic, deep pressure, pain and thermal.

14 One is tempted to say that different national traditions in philosophy diverge sharply on what they take to be the Ur-phenomena of the knowledge process, the prime starting mechanisms of the cognitive genealogy, whether sensory irritations a la Quine (in the footsteps of Hume), conatus or effort (Maine de Biran), something akin to the nature of volitions (particularly in German thought, as even the logical empiricist Reichenbach wrote of 'volitional bifurcations' in connexion with the basics of the philosophy of science), acts of faith (faith as a *prius* in relation to belief-formation), or prejudice. In a positive sense, as Gadamer argued at length as against the Enlightenment view of prejudices as begotten by sinister interests, primal beliefs or the *Ur-doxa* is when one taps into phenomenological investigations (as espoused by the works of Edmund Gustav Albrecht Husserl), imaginative spontaneity (some neo-Kantians) or whatever. In any case, even without reference to national traditions, different kinds of epistemological doctrines tend to favour or presuppose different sorts of what we might call *epistemic Ur-phenomena*, which include the varieties we have just listed, and possibly others. For Reichenbach's 'volitional bifurcations', which are rarely cited, see his book *Experience and Prediction* (1938).

15 *Tout court* is a French phrase that means 'very short', 'simply', or 'without anything else'.

16 The allusion is of course to Mauss's famous essay on the techniques of the body (which some call also the *savoirs du corps*), which introduced the concept to anthropology and the social sciences in general (1934). The contemporary 'regression of listening' was a concern of Adorno, at least from 1938, but we might speak more generally, by parity of reasoning, and in the light of similar evidence in other fields than that addressed by this author, of a 'regression of spectating'.

17 Descartes and Kant, for instance, subscribed to the topos of the 'nobility of sight'. There is a counter-tradition of Platonic inspiration more sympathetic to touch, grasping and handling as cognitive sources against the privileging of sight and the distance senses as a whole (Corbin 1986). Aesthetics has thematized the senses far more than epistemology, though in more recent times there is not much difference in this respect between a textbook of epistemology and a textbook of aesthetics. It is curious that in some European languages, until quite recently, 'beauty' was associated exclusively or primarily with the visual sense. Tolstoy

remarked on this concerning Russian, in the course of questioning whether music could be properly called 'beautiful' (a 'category-mistake' for him). Practising artists or art critics have sometimes investigated these matters (Clynes 1977), and of course for every sense there is a tradition of therapy especially of the mind involving that particular modality of sensation (going back to the classical world). Alternative therapeutics has been much more concerned with the senses than orthodox medicine.

18 For a recent account of research on somesthesis, see J. C. Craig and G. B. Rollman's paper on the topic under this title, which was published in *Annual Review of Psychology* in 1999. Social/sexual radicals have sometimes pressed for a kind of democracy of the senses, but more for the sake of total eroticization of the body than for an exploration of their untapped, disparaged or neglected cognitive, emotional and aesthetic possibilities within the confines of our dominant cultural schemes of sensorial hierarchy, practical and rhetorical. The distinction should be drawn between the '*sensualists*', seeking sexual gratification above all else – bohemian, 'bourgeois bohemian', or sexual radical – and the '*sensuists*', who enjoy the pleasures of sensory experience to the full, voluptuaries of all the senses available to them. In *A Natural History of the Senses*, Diane Ackerman gives the example of Helen Keller, blind, deaf and mute, as a sensuist (1990, xviii). In addition, the sensitivity to synesthesia has been explored by poets and mystics and presumably also sought after by true sensuists and those interested in sensory democracy in a Catholic fashion.

19 *Bien pensants* is a French phrase that translates literally to 'well thinking', and is used to refer to conventional or orthodox thinking.

20 Foucault had been preceded in this in a long paper by the leading French Lacanian (Lacan being, as noted in the text, another major anti-ocular exponent), J. Milner, whom he did not cite. Without stressing the visualist theme, the importance of Bentham's work on and preoccupation (indeed obsession) with the panopticon had been discussed at length by G. Himmelfarb even earlier. There are nevertheless some commonalities between Bentham and Foucault, above all perhaps their common 'hermeneutics of suspicion', as exemplified in Bentham's theories of fictions and fallacies, fictions and fallacies which he ceaselessly ascribed to 'sinister interests', mystifying if not self-deluded, including rights discourse and other Enlightenment ideological devices (Bentham 1843; Foucault 1977).

21 *De haut en bas* is a French phrase that means 'from top to bottom' and indicates condescension.

22 Martin Jay has given a rich, remarkable account of this French tradition in *Downcast Eyes* (1993). This tradition perhaps runs deep, for I have found an even earlier instance of this in the writings of Dom Deschamps, the eighteenth century utopian communist. It does not appear that Jay's excellent work has had the impact it deserves on students of Foucault and on Foucault-intoxicated sociology in general. How far is this theoretical 'ocularphobia' particularly salient in France? This is something that is hard to determine without further inquiry, and the topic is receiving attention by students of comparative literature. In German thought two important traditions could be so characterized: the exaltation of music not only as the highest of the arts and the supreme language of feeling, but as disclosing the depths of Being, in German philosophy and German culture at large, and of the Word in the hermeneutic tradition, theological (initially God's spoken words), literary and later philosophical and ontological. Taking into account the import of the Jewish prohibition of graven images may have been a contributory factor to *Bilderverbot* in some cases. In Britain one thinks immediately, in this connexion, of Coleridge's famous excoriation of 'the despotism of the eye'. Even someone like Herbert Spencer professed a belief in the supremacy of music in the scale of the arts and of its paramount value as the language of the emotions, which the evolutionary future would further develop, as 'the vague feelings of unexperienced felicity which music arouses. The indefinite impressions of an unknown ideal life which it calls up [. . .] may be considered as a prophecy, the fulfilment of which music itself aids' – music as a felicific self-fulfilling prophecy! (1950, 75). Poets of course have had a stake in privileging the Word as against the over-emphasis on the visual.

23 The *survol* theme however goes back well beyond Descartes to prophets and seers: 'not only is the seer removed from the *place* he normally occupies on Earth, but in the order of *time* also – and this its special feature – he is removed backwards to a period so remote that he is able thence to look forward over the whole expanse of history [. . .] he gains an understanding of the events in a perspective which makes known to him their relation to the past and the future' (Hare 1923, 308). Some read the Elias account of the process of civilization in terms of the dominance of the visual sense, in accord with the French anti-ocular tradition, but in fact in the text of Elias's first major work it is quite clear that the formation of controls pertaining to other senses were just as important and, subsequently, Alain Corbin has shown the importance of the lowering of the thresholds of smell tolerance in the ordering of urban life and bourgeois manners, although with significant cross-national variations, from the eighteenth century onwards. 'Sanitary reform', which certainly included dealing with offences to the olfactory sense, was a major concern in early Victorian England.

24 For the views of a real blind philosopher, see the text *On Blindness* (1995), by Bryan Magee and Martin Milligan.

25 In another paper, titled 'On the Status and Role of Instrumental Images in Contemporary Science' (2014), I call the fashionable position of the Imagologists, who see us today as totally immersed in an ocean of electronic images, and irremediably cut off from any other sort of reality, appearances or virtualities, as a kind of *Absolute Imagism* by parity of reasoning and denomination with Francis Herbert Bradley's position. It is not surprising that books have appeared recently comparing the thought of Bradley and Alfred North Whitehead (McHenry 1992) and Bradley and William James (Sprigge 1993).

26 This was the argument of Milic Čapek in his erudite and indispensable *The Philosophical Impact of Contemporary Physics* (1961).

27 Perhaps we may draw on Kenneth Burke again: '[W]e have come to think of ourselves as *listening* to the universe, as waiting to see what it will prove to us, we have psychotically made the corresponding readjustment of assuming that the universe will abide by our rules of discussion and give us its revelations in a cogent manner' (1954, 99; italics in the original). By 'psychotic' the author meant nothing psychiatric, just a 'mind-set'.

28 See C. D. Broad's *The Mind and its Place in Nature* (1925), which is still useful for a wealth of analytical distinctions and classifications of philosophical positions.

29 *Ab initio* is a Latin phrase that means 'from the beginning'.

30 For Sartre, pursuing a similar line of thought, and in accordance with standard phenomenology, consciousness is always 'thetic'.

31 One would have to ensure, too, that the reverse move, from verbs to nouns, was not made in a serious ontological grammar. It was said of late nineteenth-century physicists that while they were unable to detect anything but vibrations, they still felt compelled to postulate the nominative of the verb 'to vibrate' as the entity, 'the aether' filling, and underlying, the vibrating cosmos. It took a very great deal to extirpate the nominative of that verb from the conceptual repertoire of physics.

32 Many thinkers from Nietzsche to Russell have traced the key Western metaphysical obsessions, chiefly that with substances and their permanence and explanatory role, to Indo-European language structures (though Nietzsche pointed out that in the material respects Indo-European and Semitic languages are not substantially different). Accordingly, the subject–verb–object schema tended to be replicated in the very language of reason, in the Aristotelian subject–predicate logic, and the substance–accident ontology it projected, an ontology that eventually permeated Christian theology (with sporadic attempts to overcome this). Russell turned to an ontology (to which he adhered to the end of his life) in which the basic world-stuff is made up of *events* rather than substances, substrata, entities, corpuscles or other hard, lumpy stuff or any standing 'things', and the particles of microphysics are thereby construed as patterns of events. I suppose this evential ontology must count as 'revisionary' or 'prescriptive' metaphysics' in

Strawson's terms, as much as Whitehead's. Whitehead, although associated with cultural materialism, was not a materialist in the nineteenth-century sense, but an energeticist believing not only in the central role of energy transactions and the struggle for energy in the shaping of human societies (hence, the stress on technological, economic, demographic and ecological factors), but in its constitutive status in physical cosmology so that the basic reality was a Heraclitean one of fluxes. His linguistic proposals were thus inspired by his version of scientific process metaphysics.

33 In a recent, and excellent, dictionary of philosophy, the entry on dispositions starts with Ryle, the implication being that it was his conceptual innovation. Not so. To take only proximate precedents, we should mention Broad's lengthy discussion of the contrast between dispositions and occurrents and Carnap's later deployment in his paper 'Testability and Meaning' (1936).

34 A political thinker who later became a professional politician in the Labour Party tried to spell out the implications of the Rylean account of the mind in conjunction with arguments from the process of civilization theory of Elias for a critique of contemporary liberal politics; this was an astonishingly early discussion of Elias in English, as British sociologists would not take him up for several decades! (Gordon Walker 1951). The topic of the civilizational import of concepts of the mind and indeed of the human being is a singularly neglected one of late. It may be noted that the current British prime minister has acknowledged a considerable debt to the philosopher John Macmurray, who was certainly an exponent of the 'holism of the human', as evident in his book *The Self as Agent* (1957).

35 The expression 'adverbial materialism' was invented, I believe, by the philosopher James W. Cornman, who espoused this approach. But a functional theory of the self and the mind could also be characterized as *adverbial mentalism*, for the conception of the autonomy of the mental need not call for a substantial ontology, and by the same token we might also speak of *adverbial humanism* in the case of an account of the total human being in terms of powers and functions. Renouvier's account of the mind and indeed of the whole human being in terms of functions in his concept of the total human function, which I have cited earlier, or Risiero Frondizi's functional in contradistinction to a substantival or substantialist theory of the self could also be couched in adverbial terms.

36 The work of Bas van Fraassen in the philosophy of science embodies a kind of molar behaviour stance or, so to speak, a molar epistemology, inasmuch as his criterion of empirical adequacy corresponds essentially to what we can observe with the unaided senses (the mesocosm or what we might call the mesoscopic world of middle-sized objects). He has been strongly and almost unanimously criticized for this, for most philosophers of science refuse to see any essential difference between the deliverances of the unaided senses and what we can elicit from the most sophisticated detection instruments to tap the phenomena and entities of microscopic or megaloscopic worlds. The epistemological privileging of the mesocosm, of the middle-sized things we are most familiar with, does seem viciously anthropocentric (Fraassen 1980; cf. his most recent presentation of his world-view in *The Empirical Stance* 2002). Despite these standard criticisms of 'constructive empiricism', in the end even a scientific realist who believes science aims at the truth about the nature of things must agree that we are committed to what one might call a *minimal epistemic anthropocentrism* or *mesocosmism* inasmuch as in the final analysis, despite all our trust in instrumentation to plumb the depths of the microworld and the far reaches of the megaworld, our elaboration of 'non-instantial concepts' without counterparts in sense-experience, and postulation of unobservables, we have to rely on our unaided senses, which provide 'an upper limit of epistemic reliability' (Brown 1990). The apt phrase 'non-instantial concepts' is from J. O. Wisdom.

37 There is at least one large (quite a bit larger in the second edition) *Dictionary of Dissenting Economists* (2001) (though it should be said that a number of the entries pertain to scholars who also appear in mainstream histories of economic thought or analysis, even as leading figures, such as Keynes and Schumpeter). Remarkably, this pertains to the discipline in which most approaches have had general if not uncontested consent, the status of a uniquely monoparadigmatic 'science' throughout its history, and which increasingly sustains itself with

the paraphernalia of a mathematical science (an anomaly if we take 'paradigm' as it was formulated by Kuhn). Nevertheless, what other discipline in the social sciences can boast a similarly serious reference work? Yet dissident thought past and present is as much in jeopardy in, say, sociology, not to speak of 'cultural studies' as anywhere, as can be seen from the extraordinarily narrow range of authors and publications referred to in publications and bibliographies in the English-speaking world, a most meagre diet of endless recycling and epicycling despite the variety of 'schools' (truly a minor branch of the celebrity industry, but then the contemporary American and soon the Western university will revolve around professorial stars and celebs). On cultural studies, see P. Osborne's *Philosophy and Cultural Theory* (2000).

38 The French phrase *égoisme à deux* means 'dual selfishness'.

39 Newspaper reports indicate that some IVF clinics allow the same sperm donor to father up to 25 children (and, in some cases, this upper limit seems to have been exceeded by a significant, even substantial margin). These donors are therefore accorded the privilege of what one might call 'genetic sultans', and their genes accorded, via clinical decisions, special Darwinian advantages not given to normal genomes in Western societies. One of the criteria of Darwinian fitness today seems to be donating sperm to IVF clinics. Given the prevailing anonymity rules, this new ensemble of 'genetic sultans' is likely to remain an 'abstract collective' (recent court cases in the UK may portend a change). 'Genetic sultanism' (as we might put it, for in the West today it is a rare privilege indeed) is certainly an unforeseen and unwanted consequence of the demands for reproductive rights and choice implemented through the sperm market or the market for genetic goods and services at large.

Notably the UK affords a case of generalized altruism that some neo-liberals find hard to swallow, though it is impossible to believe that, in general, a market would be more 'efficient'. For the classic discussion of the significance of this practice of voluntary blood-giving without charge for the use of anonymous others (not just for relatives), making up a distinct as yet largely non-commodified sector of life see Richard Titmuss's *The Gift Relationship: From Human Blood to Social Policy* (1970).

Marriage for Kant (who of course never married) was a contract, between a man and a woman, to enforce exclusive mutual use of each other's genitalia.

40 Recent books of applied economics include one or more chapters on the economics of genetic services, genetic insurance, organ transplants and the like (Tabarrok 2002). Clearly what I have called body economics and geno-economics are up and coming fields.

41 I was told a few years ago that editors of philosophical journals were restricting publication of Rawls commentaries as otherwise their journals would be filled with nothing else for several years! *Si non è vero.*

42 Venezuela once had a Ministry of Intelligence. Such ministries of gene-directed change could go further.

43 J. Passmore (1970) had already devoted a few pages to what he called 'genetic perfectibilism' from Plato to the present. The present conjuncture involves a specific new stage of this ancient Plato-descended mode of perfectibilism with genetic science and technology, and hence I have adopted this expression. Also, it is no longer a question of political control of heredity through dictation of who can breed and with whom, but rather what one might call market enablement of directed genetic change.

There is now the prospect of a serious competitor (though there might be complementarities). 'Regenerative medicine', which through the regrowing of fresh versions of our organs as we age, holds out the promise of much increased life spans say of up to 100 to 120 years or even more, while enjoying good health and high-grade mental capacities and therefore the ability to work and study. However, in the foreseeable future this kind of medicine would be so limited by the ability to pay for the required interventions as to be accessible only to a minority, even in the West. In the first instance, it might be largely restricted to a new medically privileged macrobiotic caste, before a more egalitarian regenerative-medicine sustained Macrobiotic Society or a Macrobiotic Medical State would emerge. The consequences would be extraordinary, for as

Comte had already pointed out, changing the average life span of generations in a substantial way would alter the fundamental parameters of human existence. There are already fears of a biological divide or apartheid world-wide, as Vandana Shiva puts it, between the rich and the poor countries, between the 'lives' and the 'live-nots', between shall we say the macrobiotic and the microbiotic societies, even with the rise in living standards and life expectancy that may accrue to the poorer countries. In a way, this already obtains in the wide gap in life expectancy between rich and poor countries, between life expectancies in the order of the 70s and in the order of the 30s, accentuated by the HIV pandemic. What the technological somatophobes think of this is not yet clear. Is their way so much better or more realistic or likely to come about earlier? Will the technological somatophobes quarrel with the technological macrobiotizers? We have to speak here of *technological macrobiotization* by which I mean significant extensions of the life span secured by genetic or other biotechnological intervention in contradistinction to and superadded upon the rise in life expectancy brought about by better nutrition, public health, sanitation, antiseptic practices, vaccination and, to a lesser extent, other medical measures in the last two hundred years or so.

44 A couple of decades ago the minister of education of the UK, a fervent anti-Keynesian, expressed strong doubts as to whether there were as yet any social sciences at all (and the prime minister of the time did say that 'there is no such thing as society, only individuals and families', so one might infer that there is nothing out there to be a science of). The then Social Science Research Council, attentive to its political masters, duly renamed itself, dropping any reference to social science, as the Economic and Social Research Council (though the disciplinary structure of non-sciences remained, with their names unchanged), and it continues to be so named to the present day. It did not commission an inquiry into the nature and possibility of social science. There are of course perfectly legitimate concerns involved. For instance, *Why Economics Is Not Yet a Science* (1983) is the title of a book edited by the economist Alfred S. Eichner. But, whether such matters lie within the purview of governments or ministers is another matter.

45 See chapter 2 of Jörgen L. Pind's *Edgar Rubin and Psychology in Denmark: Figure and Ground* (2014), which is titled 'Psychology Without a Soul'.

46 This is in the strict sense of the philosophy of mind in the nineteenth century, of terminal, end-of-the-causal-line effects, effects that are not and cannot be causes (no feedback loops for them!), strictly impotent or causally redundant outcomes of deterministic processes.

47 Similarly, there is a wide interest in the kind of psychagogic techniques classically developed in Hindu civilization, subsumed by Max Weber under the rubric of the 'systematic rationalization of magic' in contrast with the modes of rationalization that came to prevail in the organization of economic life and in every other major field in the West. The difference is that this unofficial revival of interest in psychagogic practices is taking place in the context of the systematic rationalization of life under the aegis of technoscience. To be sure, some contemporary technologies, especially in information and entertainment (infotainment, edutainment), as R. Stivers has argued at length in a recent book (1999), contribute in themselves to a magical apperception of the world. Scientists do talk for a wide public in magical ways, invoking astrology and alchemy, for instance, sometimes, as when James Watson, in the context of the Human Genome Project, spoke of our fate being not in the stars but in our genes, or when Joshua Lederberg spoke admiringly of 'algeny' (genetic alchemy) in reference to genetic engineering. The historical and functional interrelations and inter-intimations of technology and magic are truly manifold and by no means purely and necessarily negative or disjunctive.

48 I have not invented these slogans. 'Information is everything' was put forward by the sociologist Sherry Turkle, who is well-acquainted with the inroads of computer technology in psyche and personality and writes about organizations. 'Everything is information' was enunciated by the mathematician, speculative thinker, and SF writer R. Rucker. Curiously, it is rare to find their obvious conjunction, as set forth here (though the first conjunct is causal-seeming, and the second conjunct ontological). The current metaphysic of information or pan-informationalism is nicely summed up thereby in this palindromic fashion (hence pan-informationalism could

just as well be called palindromic informationalism). I address these issues in the metaphysics of information -of which this palindrome could be the presiding motto and the mantra – in my book *Experimentum mundi*.

49 Some scholars have defined materiality or physicality in terms of this criterion of exclusive occupancy, though this breaks down in microphysics. At least in metaphysics, a rough characterization of a universal is that which can be in two places at the same time; nominalists require a world where this does not obtain, a world with only particulars in it, though it is not clear what they can make of the microphysical perplexities on this matter. Nevertheless, it is still tempting to regard space of the two forms of sense-intuition distinguished by Kant as the most physical or physicalizing, as it were, even as the only genuine form of sense-intuition in contradistinction to the categories of the understanding (as the neo-Kantian philosopher F. Pillon did). One of the things that most struck Lévy-Bruhl in his famous account of at least some kinds of 'primitive' discourse was the apparent ease with which the possibility of a body, a person, any object being in two places at the same time was entertained, thus leading him to his 'law of mystical participation' (1923). This 'law', of course, represented, for this much-maligned author, not just a stage of cognitive evolution but a permanent stratum of the human mind.

50 This is not to be confused with the 'logocentric predicament', an expression invented by the American philosopher J. Lowenberg some decades ago in order to indicate how attacks on logic, reason and rationality necessarily involve arguments and claims of a putatively logical or rational kind if they were to carry any weight with the nonconverted; some of the Greek sceptics had avoided speech altogether to escape this predicament, to avoid even expressing any beliefs, even sceptical ones, resorting to just wagging their fingers. Of course, 'logocentrism' has acquired a quite different but not more justifiable or illuminating sense in recent years.

51 Pierre Lévy is the supreme exponent here and a most engaging one.

52 The classical formulation of the concept and critique of the over-socialized image of man is due to the sociologist Dennis Wrong. Also, at least some social constructivists admit to ontological realism, that is, the view that there is indeed a world, a reality, independent of our minds, not constituted by and immune to our social constructions. However, this seems a totally idle claim, as they cannot profess to know this to be the case or as something that can be demonstrated to be implicated in all our conceptual life. Perhaps this is a case of an indulgence, for once, in 'phatic communion' with those they otherwise sharply disagree with, as is notoriously the case. Moreover – and this is the crucial point – such a piously affirmed reality, in their perspective, is something like the ontological counterpart of the *deus otiosus* in some religious conceptions. It may be out there, but it does not do anything to us or for us. It plays no active causal part that we can ever know, or at any rate it plays no *criterial* role whatsoever in our knowledge (in the sense of providing the truth-makers for our notions). It does not arbitrate amongst rival theories or understandings of it (for the constructivist it would be sheer mystification to talk in this fashion), just as the *deus otiosus* plays no part in our salvation history or our moral lives, not 'making for righteousness', being in effect redundant for anything that matters in the human world, basically having no bearing on any of our decisions and choices or on the human drama as a whole; the Epicurean gods were somewhat like that. Theirs is perhaps the ontology of the 'otiose reality' ('*realitas otiosa*') that does not 'make for truth', as a genuine providence 'makes for righteousness', a truly gratuitous or vacuous ontology, an *otiose ontological 'realism'* just as the theology of the *deus otiosus* is a gratuitous or vacuous theology. To all intents and purposes, they are at least 'methodological atheists' of reality.

53 *Principium individuationis* is a Latin phrase that means 'principle of individuation'.

54 This is affirmed in one of the pregnant footnotes of the last book Durkheim published in his lifetime, *Les formes elementaires de la vie religieuse* (1912). This is a key source for the charge of 'Averroism', recurrently levelled at Durkheim since the 1920s at least (Max Scheler being one of the critics). Averroism is that species of theological and philosophical heresy In Latin Christianity (Left Aristotelianism Ernst Bloch called it) which affirms the species-wide unicity

and immortality of the 'active intellect' in which all humans participate qua knowers, underestimating the role of reason in each knower or the autonomy of the knowing subject. The recent discussions of the relationship between world 3 and world 2 in the Popperian 'epistemology without a knowing subject' resemble this old problematic. Still the analogy between *conscience collective* and the 'active intellect' does not hold strictly in the case of organic solidarity, as becomes clear from the lectures on pragmatism and sociology.

55 See Elliott Sober and David Sloan Wilson's *Unto Others: The Evolution and Psychology of Unselfish Behavior* (1998). This is perhaps the first systematic examination of the concept of group-selection in evolutionary biology, its legitimate scope and warrants, taking into account the most stringent criticisms.

56 *Kulturträger* is a German term that refers to the passing on of culture.

57 There has always been at least since the late eighteenth century a special interest in the 'body electric' or the body electromagnetic, particularly on the part of writers and artists, such as Mary Shelley (*Frankenstein* 1818) and Villiers de l'Isle Adam being classical landmarks on a clear line of descent from Mesmerism, an exceedingly influential para-scientific cultural movement in its time.

58 Bernal was followed in this scenario by the vision of the Ultra-Intelligent Machine, I. J. Good, in the 1960s, and since then by many others, most recently by W. H. Calvin.

59 The concept and indeed the frank and full post-human eschatology of the Ultra-Intelligent Machine were advanced as far back as 1965 by the British mathematician I. J. Good (at the time located in Oxford) in his lengthy technical paper, titled 'Speculations concerning the first ultraintelligent machine', and reiterated in a number of other shorter presentations. Bernal, who in some ways foreshadowed this scenario, was a Cambridge man. Good does not cite him in this text though he invited him to contribute to his edited collection *The Scientist Speculates* (1962).

60 The ongoing scientific movement of 'Artificial Ethology', as it calls itself (which might also be called Artificial Animal Intelligence, A.A.A., or the nonhuman, zoological branch of AI) aims to develop animal-level intelligence and other animal sensory-motor capabilities in robots, ostensibly for the sake of a more sophisticated zoology, or advancing beyond the well-established natural history type 'cognitive ethology'. Thus, not only do we envisage our own technological obsolescence as a feasible project if not, indeed, as a necessity given the prospects of global climate change, we also appear to aim at making animals equally redundant, or at any rate equipping ourselves with the potential for bringing about this goal, at least in principle (though doubtless a number of zoophilic reasons will be advanced ad hoc if required). Perhaps it will be a close race between those engaged in Artificial Zoology or Ethology and in human-oriented Artificial Intelligence to see who comes first in the technogenesis of intelligent artificial beings, the Artificial Animal or the Artificial Person, the zooid or the android. Oddly enough, the second project, at least in an elaborate computational way, seems to have preceded the possibly more feasible or the prima facie less exigent animal intelligence simulating and attaining project. Of course, in non-computational, merely mechanical terms, animal robotics was in some ways far advanced with Vaucanson in the eighteenth century (such as his ingesting and defecating ducks) which, however, remained as *curiosa*, amongst many others in the long history of automata, not generating a continuous tradition of invention.

61 I. J. Good, in his project for an Ultra-Intelligent Machine (1965), had already posited that what he called 'ultraparallel' and distributed processing would have to come into operation before the 'ultraintelligent' being could emerge.

62 'Enfleshment' is the vernacular word suggested by some British theological writers instead of the received time-hallowed Latinate term 'incarnation' for the working Christian theological vocabulary in plain English. This explains the non-exclusive adoption of 'disenfleshment' for 'discarnation'.

Chapter Five

WHEN UNIVERSITIES BECOME BODY-SHOPS

Sorry: your soul has just died.

Car bumper sticker

News from the rat race: the rats have won!

Tom Wolfe

The recent trend, by both Labour and Conservative governments, to substantially increase university tuition fees in the UK does not spring solely from the need to address the financial crisis in higher education. It is driven also by the sense that British universities, or at least some of them, must move towards a US exemplar/myth/utopia of the 'world-class research university', or some version of it, though no clear specification of the goal-state or even of a spectrum of scenarios, appears to have been published as yet. I am not sure whether many, or indeed any, of the distinguished academic backers and co-instigators of this drive share the brutal judgment expressed recently by a former Labour minister of education, resident in recent years in Cambridge, Massachusetts, that Britain does not currently possess a single 'world-class' university or multiversity, Britain having presumably slipped down into this outer darkness at some oddly undisclosed point in the recent, or perhaps not-so-recent, past.[1] In fact, a ranking of 500 world universities and 100 European universities, prepared by a team at the Shiao Jong University of Shanghai, shows that Britain, as of 2003, was doing very well indeed in the number of universities in fairly high places on the list, in having two universities in the top ten (so part of the *la crème de la crème*), and four in the top twenty.

My concern here is not with the question of the comprehensiveness and equity of access to universities supposedly ensured by the new financial arrangements, important as it is, or with the 'output' so unengagingly described by the *Financial Times* in commending these proposals editorially, as nothing more than improved 'intellectual skills of the workforce' (2004).[2] For which purpose, surely, you do not really need universities at all, let alone 'elite universities', as the FT calls them, and it is worth noting that this proverbial 'mouthpiece of capitalism' eschews any additional reference to such *desiderata*,[3] if not sheer requisites, of a healthy democracy as a well-educated citizenry. Indeed, the FT does not at all refer to the 'citizenry', or to the 'nation', or to the 'people' of Britain, but only to the 'workforce'; it seems to imply that the sole matter of concern is the transformation of the studentry into suitable labour market material or the vector studentry→workforce.[4] It altogether fails to invoke the word 'education'. Indeed, this 'world-class' newspaper failed utterly to mention even any cognate terms such as 'culture', 'cultivation', 'civilization', 'citizenship' (national, European or ecumenic), 'formation', 'competences', 'qualities of mind', 'intellectual qualities', 'breadth of understanding', or even, unbelievably, 'knowledge', all keywords belonging for the last two centuries to the discourse of and about the university, about higher learning, everywhere in the West.

How can we account for the omission of these terms in the FT or in the UK government's statements about the role or the 'mission' of the universities or of the expectations they entertain about students? And how far is this attenuation of the governmental view of the universities representative of recent international trends? 'The City's house journal' did not mention, either 'democratic citizenship' or some democracy-related facet of education, which would almost certainly have been mentioned, most likely even stressed, on a comparable occasion by, say, the *Wall Street Journal*. But in any case, my focus is, rather, on the character of the institutions students are going to have access to, and what kind of 'form of life', what kind of form of academic life, they will be participating in, as well as their teachers.

The academic advocates of the course referred to are undoubtedly extraordinarily busy people at any time, and the exertion of political pressure, not least on our rulers, is, I am sure, particularly draining. So, it is not clear whether they have had the opportunity to read, or re-read, anything of an analytical rather than merely encomiastic kind on American universities. It is American private universities – and largely a subset of these, in effect – they unceasingly refer to, choosing to ignore other forms of excellence in American higher education, for instance the liberal arts colleges or even the great state universities. I have in mind a couple of works in particular. The first is a classic analysis of the American research university in the 1960s, set forth in the book by the sociologists Talcott Parsons and George Platt, and written with the assistance of Neil Smelser, called *The American University* (1973) (the theoretical apparatus being due to Parsons), in part indeed something of a celebration, in many ways a perfectly justified celebration, of the American university at its best, at a time when it already had a major world impact. It was the prime stimulus to the constitution of an academic world, or at least a transatlantic, or North Atlantic, market in the postwar era, not just in terms of students, for such may be said to exist since the late nineteenth century in EuroAmerica, but in terms also of researchers of all kinds and faculty members, at any rate in vastly greater numbers than earlier transatlantic or transpacific flows. Whatever the misgivings about the theoretical framework deployed, today regarded as outmoded, though not superseded as a general orientation, the general thrust of the argument was by no means controversial at the time, and it has been influential.

More accessible and more widely influential, though not embedded in a systematic analysis of the American university, and with a more skeletal theoretical apparatus, was the promotion by another Harvard sociologist, Daniel Bell, of the concept of the post-industrial society. This was his particular version of essentially the knowledge economy, in which the university, understood as a site of production of basic *theoretical* knowledge vital to the progress of high technology, as well as of more directly relevant or applied knowledge and technologies, comes to occupy an unprecedentedly central place as a key institution of economy, defence, culture and society in the West (Bell 1973). This vision was supported by the independent contemporary development of the concepts and statistical documentation thereof, such as the works on the 'knowledge industry' (by the economist F. Machlup) and the 'knowledgeable society' (by the sociologist R. Lane) published in the 1960s, or on the 'information economy' (by the cyberscientist M. Porat) in the 1970s, reflecting on the rise, spread, acceleration and interlinking or cyberlinking via digital interfaces of the new technologies of information, computation, communication, control, and automation.[5] This has implicated the spread, or rather, the saturation of the world, with ever more numerous (even superabundant), versatile, powerful, ever-faster (on nano-scales and beyond), ever more interconnected, inter-translatable and 'intelligent' machines, networks and systems, spanning the globe, with microchips embedded in all new artefacts, big and small, industrial or domestic, so that 'unintelligent' tools or machines will become, increasingly, endangered species, as well as in natural, organic beings (genes, organs, brains, under the skin), according to the imperative of universal cyborgification.[6] Such machines permeate knowledge-work and the mode of scientific and technological knowledge-production itself. We will note later the emergence of the 'robot-scientist' as it was baptized recently by the octet of non-robotic scientists who designed the first effective members of this new taxon, as well as the ever more knowledge-dependent, knowledge-intensive, technology-intensive advance in every field of production, transport, communications, media, weapons systems, 'intelligence', security, biomedicine, commerce, finance and the professions, and so forth. This material advance is often charmingly characterized as 'dematerialization', or moving towards a 'light', or even 'weightless', economy, with unlimited positive feedback loops. It should be added that Bell confidently envisaged, and took for granted, an enduring, considerable role for an enlightened state (a benign, social-reforming state) in the promotion of the knowledge university, the knowledge economy, and a more equitable society, far beyond what the marketeers would countenance today.[7] As far as this area is concerned, his vision appears wholly at variance with the current state-backed *market redesign of universities* and the overall ideology informing it.

These works, to be sure, were written at a time of unprecedented bonanza, a time that is still being called a 'golden age' for the universities – and of course above all for the research universities. It was likewise the golden era for scientific research, the fostering of which was due above all to federal government largesse, in what we can now see as the *epoch of the governmentalization* (though especially 'federalization') of science funding, which lasted from the 1940s roughly until the 1980s, and saw the surge of Big Science. The first Reagan administration can be viewed as a turning point, inaugurating the *epoch of marketization* of the sciences and especially the universities, and the increasing privatization of academic research and scientific research in general. This happened a decade before the end of the Cold War, and through the Star Wars venture, it should be noted. However, and this is a critical point, substantial federal funding for scientific research has been maintained, if on a much-reduced scale, especially in the area of health or health-related sciences, as well as of course in fields of perceived military relevance or national security. A great deal of private R&D, not least in the pharmaceutical field, is in effect parasitic on public science, and simply does not match the public-funded effort both in financial terms, and in the importance and scale of innovation. In addition to the turn in general orientation regarding the funding of science, marked by this two-term presidency, an extremely important measure in bringing into full effect the marketization of science was the Bayh-Dole Act, enacted in 1980, which afforded key incentives to the marketization of academic research, and more generally of the universities, coinciding as it did with the surge in genetic research.

Such approaches should be complemented by a work first published in 1918, Thorstein Veblen's *The Higher Learning in America: A Memorandum on The Conduct of Universities by Business Men*. This book might well have been regarded until recently as a dated, eminently forgettable tract for the times, or rather, against the times; President Coolidge's notorious utterance, 'America's business is business', came a few years later. It has much in common with the novel *The Goose-Step* (1923), by the muckraking writer and radical activist Upton Sinclair, which in a way fictionalized Veblen's vision, attacking the control or attempted control of universities (in terms of curricula, teaching and research priorities, appointments, promotions, etc.) by 'captains of industry' (or of finance) via the boards of trustees, though both reserved special venom for the megalomaniac university presidents of the time.[8] It was all before the great surge of American universities after World War II, boosted by factors such as the historic G.I. Bill,[9] and later the 'Sputnik effect', that is, the political impact of the USSR's launching of the first space satellite known worldwide as 'Sputnik [I]', which triggered the National Defense Education Act of the 1950s, further enhanced by the example of California's Master Plan for Higher Education issued in 1960, to unprecedented heights of student numbers, opulence, prestige, achievement and worldwide influence, unprecedented in America and in the world. It might therefore be dismissed as merely of historic interest, but the work regained a surprising lease on life, of relevance, of topicality, in the world of the 1980s, and subsequently, the world in which 'business canons', the 'pecuniary ethos', and business interests have impinged more and more on university life and even its self-image and public discourse. Veblen's numerous examples of business involvement, influence and attempted control in practically every sphere of academic affairs in the United States of his time, though American science, by some measures, had already matched if not surpassed both England and France by 1910. Yet many can now be matched by very recent examples, even in Britain, some rather distasteful, to say the least, such as the Murdoch Chair of Communications at Oxford, others in some ways even more sinister, as in the role of tobacco firms and other commercial sponsors in all too interested sponsorship of research and publications (with the tacit possibilities of bias or self-censorship), contribute to a diffuse but far from admirable moral climate.[10] Even if direct business control of curricula and appointments, many examples of which Veblen cited, does not appear to be such a prominent feature of the academic landscape as yet, they are by no means unknown.[11] Government promotion of certain kinds of research and development has now gone far beyond defence concerns. In particular, especially in the UK, they foster and protect biotechnology (including the more scientifically and environmentally controversial phases). They are controversial within the United States as well, as the 'scientific state' (the science-promoting state, for reasons of power, prestige and international competitiveness), has now effectively become the 'scientific-merchant state', the kind of state that promotes science-technology, or 'intellectual capital', as one of its paramount objectives, as big business. Needless to say, much of it is devoted to hi-tech weapons systems, bearing in mind that the world market for arms boomed until recently and has by no means disappeared. It might even be called the 'biotech-merchant-state', if not the 'Frankenfood merchant state' (the UK may even surpass the United States in this respect), steadfastly pursuing a course against public opinion (so much for the alleged guidance of 'focus groups'), and indeed at least a section of world scientific opinion.[12]

In any case, if one conjoins the approaches of the two works, for Veblen's text acquired an unforeseen new topicality, especially from the 1980s onwards, one might begin to work out an analysis of what one might call 'the research-business university', a locution that is less euphemistic than that of the 'research university', which should no longer be used to mislead anyone. If 'market-research university' might be confusing, at least for the time being, to designate the emerging form of life in the epochal and too-long delayed transition to full-fledged academic capitalism (or academic 'turbo-capitalism') against the remnants of academic feudalism and statism (ecclesiasticism being long gone in the UK, except in quaint vestiges), 'research marketing university', if not '*market-driven research university*', or even '*market-driven research-intensive university*' might well do.[13] The research university, without the defence contracts (by the Army, the Navy, the Air Force, and the US Department of Defense) or business contracts or the expectations of granting agencies, and the ever closer overlapping and interpenetration of science, technology, industry and market economy, with the 'new mode [or Mode II] of scientific knowledge production', simply does not exist, if indeed it ever did, except for some universities for some of the time, and not since World War II.[14] The Department of Defense itself sponsored the notorious Camelot Project on the sources of insurgency in the Third World, eventually cancelled by President Johnson, which had been envisaged by its academic leader (by no means a classic right-winger) as 'the Manhattan Project of the social/behavioural sciences'. Just as three decades later the US nonmilitary-financed Human Genome Project was envisaged equally grandly by its champions as 'the Manhattan Project of the life-sciences'.[15] One may recall President Eisenhower's 1961 Farewell Address, a text full of enduring wisdom, a historic, if neglected, document in the sociology of science, with its surprisingly cogent and still all too pertinent comments not only on the dangers of the 'military–industrial complex' but also on the potentially harmful tie up between science and business, academia and industry, higher learning and commerce.[16] It was from this speech, not from rabid left-wing tracts, that the expression 'military–industrial complex' entered general currency, though 'military–industrial–academic' (coined later by Senator Fulbright) or 'military–industrial–scientific', or 'academic–industrial', or, more inclusively, the so-called 'triple helix' of industry–universities–state (though especially the military concerns of the state), better represented the realities then and now. Though Daniel Bell wrote extensively on 'the cultural contradictions of (contemporary) capitalism', a term he coined,[17] he did not systematically address the cultural contradictions of the capitalist knowledge economy as such – in which the production of knowledge or, more broadly, 'information', as a 'public good', indeed the public cultural good par excellence, as 'gifts' to the 'global knowledge commons' (Dasgupta, Mäler and Vercelli 1997), is uneasily associated with the competitive national state and corporation pressures on the already business-steeped and business-oriented university to reshape itself as an organization, a business organization servicing other business organizations (as well as, of course, the defence establishment), dealing above all in intellectual property and intellectual capital, knowledge-as-commodity and knowledge-as-capital, as the price of survival (though, surviving thus, it will surely die as a university).[18]

It would be wrong to fail to note that the academic advocates referred to are indeed perfectly well aware, as well they might be, of at least two kinds of major consequences for the universities of the course of action they are fighting for, and, in the main, they embrace them. I have failed to detect a single note of regret or concern about such matters so far, in public statements, at any rate.[19] Both sorts of consequences are in the same direction, towards greater inequality, with a twofold bias towards bimodality, a bias towards a *bimodal distribution of resources and rewards for universities* and a drive for a much more unequal, eventually *bimodal, distribution of resources and rewards for academics.*

A Bias towards a Bimodal Distribution of Resources and Rewards for Universities

One is greater and more glaring, a more brutal inequality in a variety of ways between universities: the cream, and the rest (if the former Labour minister whom I cited earlier is to be believed, one might say, drawing out some implications of her claim, that there is in the whole wide world of universities, the 'American cream', and the rest). The Parsons and Platt classic study of the American university had illuminated the prestige/resource stratification of universities, from the top ones, whether Ivy League or the great state universities, to the intermediate and bottom layers (and the bottom could be far down indeed). Some might regard such a development as retrograde, and certainly running counter to the tradition of British universities since World War II at least, symbolized by the nationwide system of external examiners drawn from any other university in the country for both first degrees and research degrees (which practice, I believe, never took root in the United States), a tradition that many foreign scholars, with direct experience of the system, have long regarded as one of the most attractive features of the British academic scene, and one deserving emulation. But presumably they regard this dissolution of tradition, of the British university model, as has been known for decades, this break with such admirable practices and conventions, as a price well worth paying for the greater good, or at any rate to drive up market values of a few universities in world Stock Exchange 'quotations' (literally or not), though one may still ask whether Britain on its own, in the terms in which these matters are couched, will ever have in the future the capacity to match the first four 'world-class universities' (just as it cannot afford a single Nimitz-class aircraft carrier, even though militarily it notoriously punches above its weight). No matter what 'top-up' fees are charged, barring most unlikely eventualities, such as a sudden surge of super-patriotism among the cleverest. Even if it did, such a place would still not be in America, which is a necessary part of the package as far as much of the rest of the world is concerned, though it is also true to say that many people will go on being attracted by British universities for a great variety of reasons, sometimes indeed because it is *not* America. In fact, in the light of the recent ranking of world universities, Britain has been doing exceedingly well.[20]

The drive to secure or increase the places in the ranking via full-scale marketization, in part neglects the fact that whatever the rankings, Britain can lay claim to scores and scores of 'centres of excellence', in most branches of what used to be called the higher learning, not least in the humanities, acknowledged as such throughout the world, more perhaps than most countries. There is therefore every reason to believe that it could maintain or foster anew: And would not such a constellation be more than enough?[21] Today the wisdom of the saying *le mieux est l'ennemi du bien* (if it is not, it certainly can be) seems strangely neglected, if not disparaged, in the drive for maximization of rank in mono-hierarchies, league tables, and the like, the cult of the numerical superlative. Is the damage inflicted on the British university system and its former virtues (perhaps not wholly extinct here and there), likely to be irreversible, brought about by the drive to secure world-class universities, as distinct from a plethora of world-class research centres and excellent individual scholars – worth the human, social and cultural costs? It is not clear to me that they have thought through the psychological, social and cultural implications of the new academic stratification on which they could draw once more on American experience, or indeed whether, in their frantic zeal to press their case, and have their recommended policies enacted, they would even care to stop and ponder such implications and the bearing they may have on any kind of academic equality. Is not this a case of 'reinforced dogmatism' in the Popperian sense? Not only do they reject categorically opposing views, but they can also explain perfectly well why such wrong views are held, and therefore do not need to be seriously discussed.[22] But appeals to equality per se, in any sphere of thought and practice, have ceased to carry much weight, except in very specific areas with particularly vocal lobbies, and the Tawney tradition of egalitarianism, once important in Christian social thought and the Labour movement in Britain ('Old Labour'), has been pretty well forgotten. If the most apparent class divide in Britain was linked to access to, or exclusion from, university education, the class divide in the future could be between those who have gone to so-called elite universities and the rest ('mass universities'?) to 'selective' or to 'recruiting' universities, or perhaps one should simply say, to 'upmarket' or 'downmarket' establishments. These two sorts of caste-making universities could become mostly self-perpetuating.[23] There is an English genius for caste-like stratification that may reincarnate itself in this marketized-technocratized fashion.

A Bias towards a Bimodal Distribution of Resources and Rewards for Academics

The other sort of inequalization, rather overlooked or underplayed by some commentators, who do acknowledge the increasing formalization of an elite group of universities as an inevitable outcome of the proposals, is perhaps even closer to the bone. It is inequality within universities, among academics, not just between ranks, but within them also. The advocates of reform in university finance seem to have in mind *inter alia*[24] the system of professorial superstars in the United States (with paler reflections in the UK, though perhaps only due to lack of money), in the humanities and the social sciences at least, as something absolutely required for 'excellence' – which in these cases is perhaps a surrogate for fame, citations ('impact'), and the like. In these domains, and even to compete with the United States, at least in the sense of preventing, or mitigating, the outflow of high-reputation academics, at least in some areas, to the United States, by bribing them with ever-higher salaries and perks, to stay in the UK, as public honours, and other forms of 'psychic income', once perhaps highly valued in a less monetized market-grant economy, are clearly not enough by a long way. We already knew about 'academic rents', though, curiously enough, this concept is not invoked when wide academic wage discrepancies are promoted. But the 'academic bribe' has not been the focus of much study except, sporadically, and not quite under that rubric, in connexion with the brain drain from Britain to the United States and several years ago even to Germany, in some branches of the natural sciences.[25] Again, they do not refer to the plight of the academic underclass, which seems a natural correlate of these arrangements (graduate assistants, teaching assistants, postdocs and especially 'adjuncts'), although these are known even to those with only a superficial acquaintance with the American academic world. Note that part-time (in pay if not in work) adjuncts make up *nearly half if not more, of the total higher education faculty in the United States*, and this proportion may steadily rise.[26]

'Excellence', unfortunately, is already a tarnished word, as some American academic commentators have noted, for its overuse by universities (not least allegedly 'third-rate' ones) in their glossy brochures, and for its resemblance to the kind of puffs new car commercials indulge in. The term excellence, unfortunately, was incorporated into managerialese some time ago. The 'University of Excellence', as the Canadian academic Bill Readings called it a few years ago in his *The University in Ruins* (1996), which has replaced the Humboldtian University of Culture, has become, in effect, the business-like (in every sense) university, the university as business and for business, the university saturated with market-speak in its privileged (if not the exclusive) mode of self-description and self-presentation. Some advocates of the proposed arrangements in university finance have been perfectly candid about the underlying drive. The universities, they say, are the last great 'nationalized industry'. Everything these days is an 'industry' in Britain still to be 'tackled', by which I suppose is meant privatized, or something like it, steered into state-defined marketization, 'forced to be free', even if it is not perhaps implied that universities should be put up for auction like the railways, presumably in glorious fulfilment of some magnificent, unassailable, world-conquering historical project set in motion over two or three decades ago. Such are the ways of 'mythistory' to use W. H. McNeill's expression (1986).[27] Oddly, one cannot recall when they were ever nationalized, although one can date precisely when the railways, the coal mines and the steel industry were nationalized or renationalized. Could it have been a mere *lapsus calami*,[28] confusing 'national' with 'nationalized'? What about the Armed Forces, the Monarchy, the swans and the sturgeon in England's rivers, and the Established Church? This one is definitely the result of a state takeover centuries back, at a time when the state robbed a large volume of properties from the then-existing Church, the Civil Service, the National Trust, the Inland Revenue, the National Anthem, HMG, the electoral process, even, in a way, the English weather, or other natural/cultural entities and processes that may be imputed an economic dimension.[29] However, it is arguable that it was precisely the pre-eminently denationalizing, privatizing Thatcher government that did the most to bring the universities into government control, in a sense, to nationalize them (as well as to centralize public administration) – until now.[30] It takes astounding historical ignorance and abysmal conceptual illiteracy to believe, or at least to imply, that whatever is not now private business must have been once nationalized by some mischievous, ruinous, doctrinaire political fiat. The myth of Original Nationalization serves instead of the myth of Original Sin in these secular times and, therefore, could (most fittingly) and should, be denationalized by a later, symmetrical, political fiat, this time a good, sound, 'nondoctrinaire' one – the first an instance of 'social engineering', the second not, though the criteria for such a discrimination has never been clearly and sharply deployed in an authoritative fashion.[31] It has always been one of the hallmarks of political ideology, in 'agitation' but even in 'propaganda', to privilege simplistic asymmetrical dichotomies in its preachings (or enforcement of thought constrictions), in this case the *master asymmetrical dichotomy* of state/private or state/market: the simplism immediately arising from the tacit claims that the dichotomy is strict, the terms posited as mutually exclusive and exhaustive against the lessons of the study of comparative economic systems, which bring out a far less stark, indeed much more multivalent picture of patterns of ownership or management.[32] The asymmetry hinges above all in locating positive value exclusively or almost exclusively on one side, disvalue on the other, and no *tertium datur*,[33] essentially speaking. The teachings of 'scientific liberalism'[34] may enjoy more free and open criticism than those of 'scientific socialism' but nevertheless are relentlessly enforced in public policy practically everywhere.

The trends in the business styles, the entertainment and media industries in America, and Western societies at large, particularly accentuated in the last fifteen years or so, towards winner-takes-all arrangements, where quite small, even vanishingly small, or indeed non-existent differences in talent or performance, between contestants for 'excellence' (top prizes), or even thanks to sheer flukes, secure wholly or at any rate hugely disproportionate, and indeed possibly quite fabulous in absolute terms. 'Rewards' in fame, or at least media exposure, reputation, money and perks, have already been the subject of substantial studies, even books, such as the recent, well-documented and referenced one by the American economists Robert H. Frank and Philip J. Cook (1995). Again, one fears the advocates of similar trends in academia ignore, or perhaps knowingly applaud, and perhaps expect to benefit, and may even feel entitled to sole benefit from such cornucopias and can barely wait for such a state of affairs to come about. If they succeed, the academic world, at least in the humanities and the social sciences, always the 'softest' region of the academic world as far as external ideological or political pressures on the one hand, and internal bids for power of doctrinal or methodological (or 'paradigm') zealots and academic mafiosi, on the other, are concerned, is going to become another subsector of the 'winner-takes-all society'. In some areas, it perhaps already is already beginning in the world entertainment industry, including of course under this rubric in an increasing number of sports, as the neologisms 'edutainment' and 'infotainment' indicate. It will tend to exemplify a kind of *super-Matthew effect*, and not just what one might call the simple Matthew effect, that is, 'to him [or her] that hath, more shall be given', known to Merton, on either of his accounts.[35] One should, however, never forget that the 'Gospel according to St Matthew' also reads 'to those that have not even, that which he has shall be taken away', which is a more up-to-date, post-Mertonian, turbo version of 'to him [or her] that hath, much, much more shall be given', as indeed to some extent was already occurring in certain phases of academic life where, in any case, careerist ferocity has never been absent, to be sure.[36] Though perhaps we should also bear in mind the reference to Genghis Khan that Bill Gates/Microsoft's greatest rival (and actually somewhat richer than Gates) likes to make: 'the point is not only to succeed, but to make sure that the others fail'. Now we have the '*Genghis Khan effect*', if you like, not yet properly recognized in sociology, unlike the simple Matthew effect; still, it was formulated by a leading businessman, one of the wealthiest in the world today, and given the prevalence of the business model for universities, it must have some bearing on the business university (not just the business or management school within the university), but the university as business, the university for business, the B2B university or business-to-business.[37]

A perfectly legitimate question to ask in this context is surely this: *What is all this going to do to the university as an institution?* It is astonishing to find that some of the most vociferous advocates of these changes are heads of ancient Oxbridge colleges, institutions if anything can be called such, institutions nonpareil in the world. It is as if their anti-egalitarianism and radical academic individualism (even by theoretical epistemic holists and keen partisans of the 'social construction of reality') made them forget, or renounce, or at least discount, the institutional(ist) tradition they have been entrusted with. Under such circumstances, in due course the institution will not count at all, so why bother with the preservation of such colleges?[38] Indeed, why not go all the way into the already much advertised scenario of the virtual university, and the all-inclusive if not yet sole and exclusive regime of 'e-learning', and indeed 'e-science'[39] (already making some progress), with virtual researchers and teachers, as has been envisaged with relish for over forty years by e-utopians and futurists of various descriptions, not least Marshall McLuhan, though perhaps now it will start happening?[40] Already universities, at least as far as the humanities and the social sciences are concerned, tend to embody a kind of hybrid of the knowledge-as-commodity/knowledge-as-capital industry and the celebrity industry; at the limit, they will become mere collocations of a few academic superstars shining ever-so-brightly in the world firmament, and their legions of teaching or research minions, part-timers or 'adjuncts' (as they are called in the United States), who may eventually be relegated to something approaching zero-contracts, or what one might call *academic Mcjobs*, bringing the academic labour market into line with other labour or service markets of post-industrialism, entitled at most to (short) lives of (hopefully) blameless obscurity. Such places serve as collocations undergoing constant reshuffling, assembling and disassembling, according to the latest poll of rankings, patent indices, grants, publications lists in 'peer-reviewed' journals, and measured by number and physical weight, the weight example not being facetious or utterly hypothetical, but drawn from real cases referred to in Theodore Caplow and Reece J. McGee's *The Academic Marketplace* (1958), if not wordage directly, all presumably in Anglophonia, for nothing anywhere else will count, of course.[41] Indeed, the ratings and scales may perhaps be issued every week, if not every single day. Edinburgh University already claims ten peer-reviewed research papers produced by its staff every day, and some other universities in the UK would doubtless claim even more,[42] if not every hour (Edinburgh's rating would then be less than 0.5 papers per hour), like shares in stock exchanges, and indeed one may well expect universities to be literally 'floated'. If only there were someone around today like that supreme savage satirist of ideas, Jonathan Swift, to pursue this imagery as Inc., PLCs or corporations, or even as assemblies of such firms, perhaps even down to one-person firms, the oxymoronic 'university of me' (and every single academic person a 'micro-firm', if not a PLC, or whatever),[43] quoted eventually in such places or their equivalent. If indeed this has not already occurred, at least in some fashion, and certainly needing no physical abode locatable by GPS, except perhaps with the logo 'somewhere in or over America', for the sake of prestige, or any kind of collective sentiment, collective affection or rose-tinted collective memory. To be sure, the decay of the university as an object of loyalty on the part of faculty originally was partly because of the growing, indeed overriding, importance of disciplinary and departmental attachments, but under present circumstances matters have gone much further.[44]

Yet science faculties in the UK are already contemplating low or nil fees for new science entrants, given the persistent decline in applications for science places (especially in physics and chemistry) and for mathematics at universities, a trend common, with variations, to most of the Western world. They are also gender skewed, inasmuch as few women go into physics, but match or surpass men in the intake into the biological sciences. Yet the thrust of university expansion and revamped finance is supposedly designed to enhance the technoscientific capabilities of the nation, or rather, of the 'workforce'. At this point, one can see very clearly the emergence of the contradictions of academic capitalism, for the 'effective demand' from students is not necessarily for the skills and competences that are vital for the preservation of the wider turbo-capitalism, which on the other hand is the *raison d'être*[45] of the marketized university. For, as so often, the profile of the 'effective demand' of the supposedly market-wise 'customers' of higher education does not match the functional requisites of the system, certainly as perceived by politicians. Math-phobia and fear of or aversion to science are too strong, but then where do these 'pathologies' or 'dysfunctions' stem from? No matter how much students become ever more economically rational or market-disciplined, or so it was expected, given the exigencies of fees and loans, they still do not seem to pursue the 'right' subjects (from the points of view of current industry or the current government) in anything like sufficient numbers, strangely enough. Even economics, as distinct from business studies, is significantly less favoured in the United States at undergraduate level than was the case, say, ten years ago, and some American universities have stopped teaching graduate economics in the last decade. The ever-growing mathematicization of the discipline, especially in the case of the most prestigious schools, widely regarded (even within the profession) as exorbitant, even byzantine, is undoubtedly a factor. In the UK, presumably, modern-language faculties would not be permitted to imitate the science faculties, even though they are also suffering from an adverse trend in applications as in other EU countries, even between neighbours like France and Germany, increasingly disinclined to learn each others' languages, at least at school and university. Their foreign-language skills (or knowledge of, and fluency in) could also be rated as important for business. So, what is doing well at the undergraduate level? In the United States, at least, business courses, enrolments in which as of 2000 made up almost *one-third* of the total undergraduate population. In the UK, a top-rated university department according to the state-imposed (though implemented by willing academics) Research Assessment Exercise (RAE), and therefore a jewel in the University of Excellence, may be closed down if numbers of applications from prospective students are deemed insufficient by the academic managers, having no place in the University of Excellence-as-business. There seems to be a cultural contradiction in academic capitalism when the requirements of excellence and the requirements of solvency clash, and the only academic-managerial solution is to sacrifice research excellence and cut costs by firing quality academic staff and closing top-notch departments or centres. Gordon Brown, the former chancellor of the Exchequer and prime minister, has been reported in the press as aiming at a 'nation of entrepreneurs', perhaps coextensive with the total 'workforce', or at least the 'skilled workforce', though largely consisting of managers and the managed.[46] This is the new version of the 'nation of shopkeepers', where the new Order of British Entrepreneurs(hip) will finally match or surpass the Order of the British Empire. Will there be also an Order of British Managership?[47] But today, surely, mathematical, natural-scientific and technological knowledge and understanding are required as well, not to mention the ability to speak and read modern languages, amongst other things. Perhaps foreign students taking such courses are expected to keep teachers of those subjects in employment, though the teachers themselves may have to be imported as well. Nevertheless, there may still be money to be made ('value added') from the entrepreneurship. Though even cleverer, even more entrepreneurial entrepreneurs may in due course simply outsource more and more and, at the limit, everything, abroad, and finally 'delocalize' the whole business from the UK to China, where some British university campuses have already opened (more due soon), perhaps the vanguard of a wide trend, in the wake of the TNCs. There is 'offshore academic capitalism' for you, following the car industry and sundry others. This is an example of the cultural contradictions of national academic capitalism, of the marketization of the university in the world knowledge/education economy, though doubtless some marketeers, libertarians (not least cyber-libertarians), anarcho-capitalists, cosmopolitans and globalists would not see anything wrong at all with such a trend.

The call for higher, even discretionary, university tuition fees, is being accompanied in a number of countries (for it is a worldwide trend, not confined to the UK or even to Europe), as a kind of quid pro quo, by demands that the residual Civil Service–type protections of university teachers be scrapped completely. Perhaps as part of a drastic denigration and drive towards the liquidation of Weberian-style rational–legal bureaucracies, as well as of other kinds, everywhere. Just as general labour codes have been and continue to be under attack in European welfare states, and one-year or otherwise time- and security-limited contracts substituted for the classical ones, if not worse, non-academic labour has been encompassed more and more by 'contingent' forms of employment. Why should academic labour, however intellectual, professional, high-grade, 'creative' or what have you, not follow suit?[48] This might well mean, at the asymptotic limit, that absolutely everyone in a university, except perhaps the superstars, and a privileged few in the administration, would have something like one-year, or, at any rate, short-term contracts, everyone, whether academic or other staff, students, and so forth would be at short notice, everyone in such a place de facto and de jure a transient, expecting to stay for only a short while (as has been the case in some ebullient university departments for quite some time, even in England), like football teams recruited from all over the world, whatever the country they supposedly represent, with the players always on the move through 'push' and 'pull' effects, in the language of economics. The upshot would be something like a post-modernity or supermodernity market *University of Transiency*, precariousness, indeterminacy (bluntly, with everyone in contingent forms of employment, as has been increasingly the condition of most forms of non-university labour, as well as concepts, as the deconstructionists have taught us at interminable length). This is if indeed the extant universities themselves as corporate entities do not disappear through merger and bankruptcy, partaking to some degree of the transiency of all their components, and mimicking even further the paragonic corporate world. The University of Excellence approaches ever more closely to the norms of the *University of Transiency* or, in the economorphic language more suited to the times, the *University of Maximum Throughput*, in personnel as well in other ways, such as ever-longer publications, grants and patents lists, per day, in ever-higher per capita rates (all part of the great 'treadmill of production'). The transiency attaches to the publications output, too, as something like 80 per cent of natural-scientific publications are never or rarely cited, and also to the graduate 'output', as engineers (of every description), doctors and others will have their specific knowledge base out of date within a decade or so, at current rates of 'knowledge increase', for there is no way to obviate the continual reproduction of 'trained incapacities' of the 'workforce'. That is, if one may mix Veblen's with the FT's idioms in an epoch of informational-technological acceleration to exponential and indeed faster-than-exponential ('hyperbolic', in the mathematical sense) heights. The ongoing 'knowledge explosion' is of course accompanied by an 'ignorance explosion', the obsolescence of university-educated professionals in fields of rapid technological or technoscience or technomarket-generated advance, such as practically all branches of engineering (and more and more fields, professional, as well as semi- or para-professional, are or will be affected by such changes and rates of change, biomedicine being perhaps just as striking a case as an engineering subject, and even more perhaps than engineering to pressures from marketing new products and apparatus), as documented by the engineer J. Lukasiewicz, not to mention the uncertainties provoked by the unprecedented scale and cumulative, even accelerating, adverse anthropogenic and technogenic impacts on the biosphere and the atmosphere.[49]

Notes

An earlier, much shorter, version of this chapter was published in 2004 (in English), in the electronic edition of the Portuguese journal, *Metacrítica, Revista de filosofia.*

1 The former Labour minister of education referred to is Shirley Williams.

2 Given the well-known, widely endorsed projection that today's graduates will have to pursue several (about 'five' is a widely quoted figure) consecutive careers in their occupational lifetimes (quite apart from periods of unemployment) in a world of 'flexible' labour markets and volatile capital markets, and experience ceaseless technological and organizational dislocations, constantly threatening their jobs, positions, incomes and prospects ('rents'), it would seem necessary to specify that these 'skills' will have to be widely applicable, for otherwise they will be shortly obsolescent in the prevalent conditions of rapid techno-organizational change. The

key emphasis should therefore be on the *capacity* to rapidly change or update whatever specific skills they may have acquired at any particular juncture, at university, or at the workplace, rather than 'skills' as such. One can no longer expect 'guaranteed jobs' in any sector or subsector of the economy, though this expectation is still entertained as a kind of 'cultural lag' by some regarding the IT or ITC sector, which, surely, like all others will not escape the general presumption. The best hope is to strive to secure the *capacities* for the likelihood of 'lifetime employability', which, given the disarray in social security systems owing to the demographic profile of Western societies, may turn out to be the other face of compulsory 'lifetime employment', as retirement thresholds keep being put back. But the simple monosyllable 'skill' has become a contemporary magic word, a substitute for thought.

3 *Desiderata* is a Latin term that means 'desired things'.

4 'Workforce' itself is an interesting lexeme in this context. The historical semantics of this expression does not appear to have been investigated. Formerly, indeed until quite recently, 'labour force' was more usual, in descriptive, or even in analytical, economics, while in Victorian Britain the metonym 'hands' was quite enough for the same kind of population.

5 For further references, see Fritz Machlup's *The Production and Distribution of Knowledge in the United States* (1962), Robert E. Lane's 'The Decline of Politics and Ideology in a Knowledgeable Society' (1966), and Marc Porat's *The Information Economy* (1977).

6 Some are truly useful, as it takes a machine to beat a machine: 'les téléviseurs américains et bientôt européens s'équipent, de plus en plus, de magnétoscopes intelligents, permettant d'éliminer la publicité en temps réel des programmes télévisés' (*Le Figaro économie* 2004, 8). An earlier Japanese attempt to provide this sort of service was, I believe, blocked by the vested interests. Of course, advertisers are resorting to other tactics, technical or legal to preempt this development: the show must go on! At any rate one wonders why these clever devices are not more widely known and available at this date.

7 In fact, Bell seems to have had in mind, like others at the time, the creation of an American Establishment in something like the sense of the term in relation to the UK, but with a much greater weight and saliency for scientists, technologists and academics than was ever the case in Britain. As Bell reports, those who shared this dream 'blew it', partly owing to the Vietnam War.

8 He drew attention, too, to the dismissal and blackballing of radical academics in every field, in the universities of the time (which was eventually repeated in the McCarthyite period of compulsory loyalty oaths in the 1950s, both in government service and in the universities, most in receipt of major federal grants, and the general suspicion of political dissenters). The picture he drew of the conservatism, political, religious, cultural, racial, of colleges and collegiands, of the time, was not, surely, much exaggerated, if at all. This 'muckraking' work apparently was rediscovered, forty years later, by some Berkeley students who played a salient role in the 1968 rebellion against the 'multiversity' of then president of the University of California, Clark Kerr. Perhaps, this time round, some faculty, at least, will read this once again all too apposite work. Maybe it all goes round in forty-year cycles.

9 To be sure, the early postwar expansion of American universities, although starting from a higher level, by various criteria, such as the 'participation ratio' in higher education of the most relevant age cohorts, can be seen as part of a worldwide, world-systemic process. Though there was nothing elsewhere quite like the G.I. Bill, there was significant access of ex-servicemen into the universities in other ex-belligerent countries (Meyer et al. 1977). It has been estimated that about 60 per cent of the baby boomers, that is, according to the conventional definition, the Americans born between 1945 and 1964 (some 76 million people), received some kind of college education, though their counterparts in Western Europe must have fallen far short of this proportion. Perhaps the current expansion-cum-marketization of Western universities marks another historic wave, converging towards a similar threshold in participation rates in higher education of about 40–50 per cent of the relevant age cohorts. Britain does appear now as perhaps the world's leading marketizer of universities, next to the United States, certainly

in Europe. However, it should be noted that other European countries have started moving that way, and one can hear the marketizing, managerializing, business drums, loud and clear, day in and day out, almost everywhere in and around universities in every European country, from academics as well as government ministers, often with what appears to be the zeal and lamentable one-sidedness of recent converts and the tunnel vision of fanatics. But I write from what used to be known, a long time ago, though perhaps no longer, as the 'home of lost causes'. A third wave of massification of higher education (or post-secondary education) and a new threshold of market-intensification and technological intensification of academic endeavours will occur if for profit e-universities (or 'digital diploma mills') take off and enlist the many millions envisaged in a new academic gold rush, a frontier, and maybe the Final Frontier, of Darwinian academic capitalism.

10 There is a notorious record of the suppression or distortion of research findings, even of their own commissioned research, over several decades, which malpractices have been in the public domain for some time, and still continuing, according to recent reports. Recently, Harvard Medical School has increased the amount of money its faculty can earn from companies benefiting from their research (stock in such companies, plus consulting fees and honoraria, in a grand total of $500,000). Yet it has been pointed out, on the basis of studies published in the last few years in the *American Journal of Bioethics*, that even fairly small 'gifts' may affect scientists. Perhaps the right analogy here, and certainly one well-suited to medics, would be with drinks and driving, where to be on the safe side it is best not to drink any alcohol at all beforehand, or with smoking, where strict abstention is recommended, by the self-same people or the same sort of professionals who in this context claim to be unaffected by heady 'gifts'.

Some African universities appear to have become virtual branches of multinational corporations to a degree perhaps unmatched in world history. If so, this phenomenon of extreme marketization of universities in the 'periphery' may exemplify the workings of what we could call, paraphrasing a famous Russian Marxist formula, the '*law of combined and uneven marketization*', as applied to universities, now a pretty formal part of the 'knowledge-economy' or 'knowledge-capitalism'. Obviously, it can be applied to all the types of organizations, industries and institutions to which comprehensive marketizing recipes and programmes have been applied throughout the great wave of policy-driven marketization in the Western world since the 1970s. The 'periphery' may provide more startling examples of extreme marketization in universities and in other areas, than the 'centre' (I refer again to this 'law' later in this chapter). But there are now at the 'centre', despite the collapse of the dot. com bubble, increasing numbers of for-profit online 'universities', or rather 'digital diploma mills', as the sociologist David Noble (who has endured a degree of academic persecution as a result of his critical stance) refers to them in his various studies on the subject, in the United States, combining the use of computers, websites and the Internet to provide standardized courses in 'distance education'. More insidiously, despite the continuing nonprofit commitment, there has been ever-increasing compliance with managerial techniques and business canons throughout the universities, assimilating them to the paragon of economic performance and virtue, the business corporation, as adverted in the body of the paper. A number of recent studies on American universities and colleges, such as the 2003 collection edited by Benjamin Johnson, Patrick Kavanagh and Kevin Mattson, titled *Steal this University: The Rise of the Corporate University and the Academic Labour Movement*, have provided abundant details on the 'reshaping of the university in the image of the corporation', which has been taking place particularly in the last decade or two, as well as of the unionization of the underclass of academia. Of course, not just the (not-for-profit) universities, but the entire independent nonprofit sector in civil society, as well as the nonprofit public sector, has been subjected to very substantial pressures to adopt business management procedures to professionalize fund-raising, to secure revenues through sales and fees, and to resemble more and more closely and overtly the profit sector thereby.

11 The sudden, unexplained decision of the University of Toronto to rescind the senior appointment it had offered to an internationally known British academic, David Healy, is widely believed to have had something to do with the circumstance that a major sponsor of the university's centre who had approached him was a pharmaceutical company, the academic in question having been on record as a sharp critic of some psychopharmacological testing practices and expressed concern over the widespread medical prescription of Prozac, as can be seen from his book *The Anti-Depressant Era* (1997), a fact that, of course, was well known long before the appointment. Presumably it was only the administration that had sudden second thoughts over the fact that the manufacturer of Prozac was the company sponsor in question. Eli Lilly has also acted promptly to withdraw financial support from centres that have published academic research critical of Prozac, not long before the patent lapsed anyway, in 1991. Yet there is widespread agreement outside the pharmaceutical industry, especially the largest dozen pharmaceutical companies, or 'Big Pharma', that it is due for fundamental reform, not an easy task given its economic power, academic connexions, for many biomedical or biotech academic researchers and academic centres have now a direct financial stake in this kind of business, and political influence, as one would expect from the fact alone that Americans spend 200 thousand million dollars a year on prescription drugs. For further references, see Marcia Angell's article 'The Truth about the Drug Companies' (2004b), and her 2004 book with the same title, and the sub-title *How they Deceive Us and What to Do about It.* Marcia Angell is a physician, a former editor-in-chief of *The New England Journal of Medicine*, and senior lecturer in social medicine at Harvard Medical School.

12 An advisory body that had expressed, if hesitantly, too many misgivings about GMOs, though not so many and not so strong from the standpoint of the outspoken critics outside, was put to pasture by the minister in the Labour government in 2004. Obviously only enthusiastic approval for this sort of thing will do.

13 The concept of 'turbo-capitalism' was advanced by the Washington-based scholar-consultant Edward Luttwak (not at all a Lefty) in *Turbo-Capitalism: Winners and Losers in the Global Economy* (1998). A similar concept of 'fast capitalism' was advanced independently by Ben Agger. In some Latin countries, the expression 'turbo-profs' is used for academics who teach in a number of universities or other institutions, in various parts of the country, at the same time, and thus spend a considerable amount of time on planes, trains and motorways, every week, if indeed not every day. I have in mind by academic turbo-capitalism or the turbo-capitalist academy something rather more complex.

The work by the political scientist Colin Leys, entitled *Market-Driven Politics* (2001), has substantial chapters on the marketization of the NHS and on television broadcasting in the UK. Of course, the market-drivenness in question was not simply the outcome of commercial lobbies, but imposed in the first instance by the state and often renewed by the state to keep institutions on course in this fashion, should the spontaneous social processes therein lead to the by-passing of market fixes. It is always selective, of course. To give just one striking example, out of many that could be adduced, consider how governments consistently fail to impose proper market charges on road-users, above all drivers, not by any means a financially or ecologically trivial matter. This insight could perhaps be expanded into something that might be called the *Law of Uneven Marketization*, a formulation surely instantly recognizable by the ex-Marxist marketizers (in the Lords and elsewhere), of universities, as of other areas of public life, though perhaps they might prefer the more complete formulation of the *Law of Uneven and Combined Marketization*, though that is perhaps too reminiscent of arguments over social and political breakdown, and presumably they are now at last wholly at peace with the social order in general and expansive all-out marketization in particular.

14 This of course also went to think tanks, some of which were set up in the postwar era in direct connection with branches of the armed services, like the Rand Corporation, funded by the Air Force. For a while after 1945, the Office of Naval Research was seen as the saviour of basic

science. I recall that even some of Eisenstadt's work was financed by them, though this happy conjuncture did not last long.

'Mode II' was coined by a group of sociologists of science, perhaps in desperation (Gibbons et al. 1994). Later in the text I suggest a slight emendation of this concept in the light of recent developments in the agents of scientific knowledge production and in any case the workings of the Law of Uneven and Combined Marketization as applied to the hard sciences, which have proceeded and are proceeding further even than the prescient authorial collective had envisaged. Molecular genetics has notoriously become a kind of vanguard of and for the marketization of science, well ahead of other branches of the natural sciences, and setting a standard of far-reaching commercialization and capitalization that others find difficult to emulate. But it has in any case contributed to a more commercial, industrialized, commodified and capitalistic image of science overall and has reinforced the political pressures for the marketization of all science. The 'technological a priori' imputed to science by some philosophers of technology is now reinforced by the very stringent 'market a posteriori', as we might call it, of business-orientation to an extent that market anticipation has come increasingly to play a role in the initiation, the selection and the relative staying power of particular lines of scientific research. Perhaps we should speak of a *'techno-market' a priori* of contemporary natural science.

15 It took shape in the climate of the 'war on cancer' announced by President Nixon. This has turned out to be a permanent war, not something that could be over in ten or twenty years, as was naively expected at the time by some politicians and operhaos by sections of the populace, though the aetiology of cancers in advanced industrial societies with their ever-larger over-60s populations, is only partly genetic or genomic, and environmental carcinogenic factors within our anthropogenically and technogenically transformed milieu have been increasingly diagnosed. But no multibillion-dollar biomedical–ecological–social sciences ten-year 'Human Environment Project', in succession to the now essentially completed Human Genome Project (and what was not completed is being speedily pursued further by private concerns), has been announced within the context of the 'war on cancer', which presumably is still on, though the prospects of total imminent victory through the wonders of 'algeny' or 'genetic alchemy' or other biomedical high-tech interventions no longer beckon as they did three decades ago, in this context, at any rate. Not even a Decade of the Environment, or a Decade of the Earth, comparable to the Decade of the Brain.

16 President Eisenhower's Farewell Address (1961) is available, and easily accessible, on the Internet. It is a must read.

17 See Bell's article 'The Cultural Contradictions of Capitalism' (1970), which was later incorporated into a 1976 book with the same title.

18 Top university administrators constantly call for more and more 'commercialization of research output'. If the university is a business, or aims at becoming like a business corporation, then it will presumably have to follow business imperatives, and I suppose the appropriate maxim and of the research sector could be the highest possible returns from the commercialization of 'research output', regardless of all other non-pecuniary considerations about the academic commons and the public good. University research findings are not envisaged as contributions to the knowledge commons but as 'output', and 'output' should be turned into 'products', that is, vendible commodities.

19 The zeal of the UK academic marketeers seems rather odd at this time, given the fact that several former presidents of top American universities, such as Clark Kerr, formerly president of the University of California at Berkeley, Derek Bok, former president of Harvard University, and William Bowen, former president of Princeton University, have published articles and books in the last few years expressing serious concern over what has been happening to American universities, not least the top ones. They are far from being lone voices in these matters, though presumably they should command special attention from those ill-disposed to listen to such concerns from any but the most unimpeachable sources on ideological or

professional–biographical grounds, even though there may be in fact some overlap between their strictures and those of more Left-inclined or radical commentators. Do the current UK academic marketeers (or their counterparts elsewhere) genuinely want to learn from America, or are they simply so ideologically or otherwise committed to the marketization of universities, come what may, that they choose to ignore the warnings and concerns of such people? There is still time to read Derek Bok's book *Universities in the Marketplace: The Commercialization of Higher Education* (2003), which surely cannot be ascribed any kind of anti-market ideological prepossessions or un-American bias; other relevant works include Eric Gould's *The University in a Corporate Culture* (2003).

20 'Top-up fees', 'up-front', 'skills'. So many monosyllables in talking about universities! – and mostly *deictics* too! It seems that in talking about universities only short, sharp monosyllables will do in current political language, perhaps a reversion to a kind of Basic English as a revenge on sophisticates, or a demonstration of debased populism. Political populism is always also a matter of language, diction, tone of voice, and finger-pointing. I am not sure whether this kind of deflated pop-language à propos of the universities is also being deployed by university marketizers in other countries besides the UK.

21 There is one crucial point about the proposed 'top-up fees', to be paid 'up-front' that cannot be stressed too often. It is that they would yield the equivalent of *one half of one percent* of public expenditure! The political costs incurred by the government in the parliamentary battle would hardly be warranted in rational political calculation, unless issues of far greater importance than this comparatively very modest amount of public expenditure, at a time of healthy public finances, were at stake. What drives the battle, one may hazard (and, perhaps because of sheer ignorance, I know of no better hypothesis), is not the saving of public money and taxes, to the frugal tune of one-half of 1 per cent of public expenditure, but the project of further drastic marketization of the universities, the final submission of the universities to the canons of business enterprise. The universities, it seems, must not be allowed to stand apart from the sweeping marketization, a thoroughgoing societal project that brooks no nonmarket, noncommercialized 'islands of separateness', no nonmarket 'anomalies' (unless they do not matter at all). That would run counter to the ideology, or indeed the metaphysic, of 'market monism'. Some supporters of these reforms have apparently claimed that education, or at any rate higher education, is the 'last great nationalized industry' in Britain still to be 'tackled' – that is, by government. To be (at least socially speaking), or at least to have a claim to social recognition, is to be part of the *market*, preferably a willing, even enthusiastic part of the market, from this standpoint. If you are (a willing) part of the market, you are part of the solution; if you are not, you are part of the problem (but the solution is obvious, imperative, urgent: To market! To market!).

22 The locus classicus is Popper's *The Open Society and Its Enemies* (1945).

23 Last year, 50 per cent of all research funds went to five universities in England: Cambridge, Oxford, Imperial College, University College London and King's College London. As it happens, the first four were the top British universities in the recent ranking of world universities by the Shiao Jong University of Shanghai (Institute of Higher Education), and indeed *all four were amongst the top twenty universities in the world*, respectively no. 5, no. 9., no 17, and no. 20. All other universities among the top twenty were American (Harvard at the top), with the single exception of Tokyo University. King's College London was no. 75 in the ranking. It may not matter at all to the powers that be, but these four universities are *the top four universities in Europe*, according to the same ranking, and King's College London takes a respectable no. 22 out of the 100 European universities listed. Twenty-two British universities appear in the list of 100 European universities, the largest national contingent. Cambridge is the second university in the world on the Nobel score, second only to Harvard, and far superior to all the others.

24 *Inter alia* is a Latin phrase that means 'among other things'.

25 I gather the term 'bribe' is currently used by some economists in a supposedly value-free sense, and wider than the conventional one. I do not know how this wide 'bribe' category would

appear in the appropriate mathematical presentations, now mandatory in the mainstream of the discipline, or how it would be operationalized in 'transparent' academic accounts or in the GDP of the academic knowledge–capitalist economy.

In some countries, this phenomenon is more forcefully called 'brain flight', more in line with 'capital flight' (this was also, once upon a time, known on the left as 'capital strikes'). Early in 2004 there were, according to widely publicized reports, for instance in *Time* and *Newsweek*, something like 400,000 European scientists working in the United States, the majority in universities or university-type institutions or research centres. This definitely beats, at least in numbers, all previous migrations of scholars in history, which were mostly caused by wars, revolutions, regime changes and ethnic or religious persecutions. Doubtless much reflection and inquiry is called for to account for it, in terms of some conjunction of 'pull' and 'push' factors (social scientists have generally argued in connection with voluntary migration that 'pull' is decisive), and to elicit ways to ensure that Europe benefits more directly from this vast pool of scientific talent, a kind of Grande Armée of European-educated scientific labour or, some might say, the largest, most talented and longest-lived 'brain strike' in history, or the largest diaspora of scientific talent from Europe (and indeed from all over the world) into one country. This is a world-historical *unicum*, given its scale and import. The recent publicity afforded to this large-scale phenomenon, and to the putative reasons for their decisions to emigrate, at least for some years, to the United States, may help to induce policy reforms in the exporting countries, and the unheeded complaints of the scientists who emigrated may in this fashion have a significant impact in a roundabout fashion, an instance of what may be called *vicarious voice*, to modify Hirschman's well-known schema *of exit* and *voice* (1970), wherein 'exit' appears as an indication of the failure, contingent or systemic, of 'voice', though it may of course stimulate 'voice' under some circumstances (see *Addendum 12*). There are of course some American scientists, in various branches of the natural and the social sciences, working in Europe (and if the political climate in the United States worsens, there may well be more), but the disproportion is enormous.

26 In fact, if one takes into account graduate assistants, postdocs and adjuncts engaged in college teaching, only about one-third of all college instructors in the United States are now on the tenure track, and the proportion of student credit hours taught by tenure-track faculty may be even lower (Johnson, Kavanagh and Mattson 2003). One could say, therefore, that, in the best managerial-business style of today, *teaching has been substantially 'outsourced' in American academia*. Of course, this is in some ways a modest form of outsourcing, for there is still human contact. In e-universities, outsourcing will reach new heights, dispensing with human presence altogether, though perhaps not some degree of 'telepresence'. Doubtless the UK exponents of marketized élite research universities expect no less in their home ground (in all the senses of 'expect'). Though perhaps they should beware of a wave of unionization of the academic underclass comparable to that in the United States, which has scored tiny successes here and there, though as the times are so uncongenial to unionization, perhaps no sleep need be lost by the academic powers-that-be.

27 McNeill was generalizing from F. M. Cornford's *Thucydides Mythistoricus* (1907).

28 *Lapsus calami* is a Latin phrase that means 'slip of the pen'.

29 I should stress that this is *not* an entirely facetious remark, contrary to what most people would suppose on first hearing this. There have been serious suggestions by technologists regarding the feasibility through hi-tech of partitioning sections of the atmosphere with a view to investing them realistically with property rights, by states or other political entities, and thereby securing the foundations for progress in yet another sphere, as it were. And certainly, weather forecasts for the next few days can only be secured from the national meteorological office by advance payments in France at any rate (though, under public pressure, the director ruled in July 2004 that such forecasts for France as a whole, as distinct from particular regions or areas, will henceforth be free).

An interesting English institution, the Trustee Savings Banks, which belonged to no one, was nationalized (taken over by the state), by then an unusual move, precisely *in order* to be privatized, to be sold off, to belong to someone in the market economy rather than in the public domain. Such is the dialectics of contemporary market capitalism! The provisions to allow demutualization of building societies in the UK on demand from shareholders were similar in intent, and they have mostly been demutualized under pressure from a minority of shareholders wanting fast cash from the proceeds of the conversion. Even when the pressures for privatization and commercialization are less direct and obvious, the nonprofit sector (charities, museums, zoos, etc.) outside public education is everywhere undergoing so much commercialization as to increasingly blur the difference between it and the formally for-profit organizations. This is a development that can only be welcomed by the partisans of 'market totalism'. But is it the most appropriate transformation in a *pluralistic* society? It is not clear that pluralism in the sense that it was used in political science in recent decades any longer applies to the current state of affairs in the polities of the United States or the UK.

30 It is arguable that a kind of nationalization took place under the Thatcher government, or a big step was taken toward towards such a condition, especially via the 1988 Education Reform Act, and further moves are taking place with state-determined, state-controlled marketization. Once again, nationalization was the prior step on the way to secure eventual privatization, or at any rate as much marketization as possible. The conservative journalist Simon Jenkins, who subscribes to this view, describes most British universities today as looking and feeling 'like down-at-heel nationalized industries, working part time and facing obsolescence' (2004). A. J. Grayling, a professor, laments in similar terms the sorry state of British universities as 'the crowded, peeling, tottering institutions called universities in today's Britain' (2004).

The government's very one-sided attitude to the BBC was in connexion with the 'Kelly affair'. David Kelly was the weapons expert who disagreed, it seems rightly, with the official view over the presence of weapons of mass destruction in Iraq, in line with which the UK went to war with Iraq, and committed suicide under pressure from the ministry and the media. This case may indicate its general outlook on public, independent institutions (independent from the state, independent from corporate interests, endowed with a Royal Charter). Perhaps we are moving a little closer to Orwell's Ministry of Truth or, at any rate, to a situation in which communications, news, and reporting will be exclusively a function either of government-controlled organizations (official news management), on the one hand, or of business media corporations on the other (corporate news management) which in some circumstances, even in a democracy, may converge. Whatever the rights and wrongs of the BBC reporting that led to the government initiating the Hutton inquiry, and the need for reform in some BBC procedures, the most important point here is that, regardless of any such details, the very notion of 'public service broadcasting' sits uneasily with the prevalent pan-marketizing stance. The presumption is of the superiority of market arrangements, whatever the field, if indeed it may not appear as wholly alien, incoherent, wholly unintelligible even, to the marketizers' mindset. Concern has been expressed over the political independence of the Office for National Statistics under the present government, as indeed under the Thatcher government, which ensured that certain kinds of socio-economic statistical data were not gathered or published. In the age of the knowledge explosion, of the knowledge-industrial economy, of the high valuation of intellectual property rights and intellectual capital as the supposedly necessary and inevitable instrumentality for maximizing the rate of technological advance, we witness thereby a kind of power insufficiently noted in a democratic polity: the power to produce ignorance and 'unknowledge' (as that exceptionally literate economist, G. S. Shackle, liked to put it) about the economy and society. Perhaps we might call this *nescience power*, not only the prerogative of governments, particularly in their dealing with the social sciences, and social and economic data, but also exercised, of course, by private corporations, not least those involved in technoscientific areas, the pharmaceutical companies affording a vast array of examples of the exercise of such nescience power regarding biomedical matters. Political power may concur in the exercise of this nescience power, or add its own ideological biases, as in the case of the Bush administration, a great friend of oil interests (and high-energy consumption, to which it appears committed to an unprecedented degree, and certainly beyond the levels of recent administrations) and biotechnology in any case.

31 Even under this government, the recent ministerial moves concerning the railways privatized by the previous administration (otherwise, presumably, they would have done it themselves) suggest that something like an unannounced return to some form of public ownership or control of this important sub-sector of the economy (which in any case was always subsidized by public funds, quite apart from being practically given away when privatized by an earlier government) is taking place. But of course, one cannot blame the market, only the bungling, if not corrupt, politicians and civil servants ('government failures', not 'market failures').

32 The remarkable study by Peter Wiles in *The Political Economy of Communism* (1962), even though published three decades ago, delineating ten or twelve patterns of ownership of enterprises and a variety of patterns of management, is particularly instructive (the *combinatoire* of logically possible combinations of ownership models and management models would be quite impressive). Some presumed Marxists or at least Marxologists, such as the LSE economist Lord Desai (a Labour Peer), have argued that there is no alternative to full-market capitalism of the sort that now prevails in the United States, but full socialism (not social democracy or democratic socialism) is impossible, at least for the foreseeable future. That, they say, is Marx's real lesson ('Marx's revenge'), which had not been heeded by social reformers under capitalism: give up all social democratic illusions! (which in the present conjuncture entails nothing less than: embrace full market capitalism!). With this kind of bourgeois Marxism, capitalism needs no more friends. This kind of Marxism-without-hope (that is the very definition of bourgeois Marxism) could well become a serviceable creed for the actually existing state of affairs, or indeed for universal marketization. Of course, there has always been a strand of Marxism, as of anarchism and other forms of social radicalism subscribing to the *politique du pire*, which virulently opposed bourgeois, Keynesian-style economic and social reforms on the grounds that by mitigating present pain and collective misery, it detracted from revolutionary fervour or the raising of class-consciousness.

On the role of asymmetrical dichotomies as foci of social thought and orientation, specifically in the constitution of political ideology and indeed of social theory in general, the writings of Stanislaw Ossowski on class theory (1963) and W. Baldamus on social science methodology (1966) (both sociologists coming from Marxist traditions), are particularly relevant. There is of course much on binary oppositions in structuralist literature, and subsequently in feminism, though another important tradition in which asymmetrical dichotomies play a fundamental role at a much higher level of analytical abstraction is that of Durkheim (the *Ur*-dichotomy here being that of the sacred and profane) and Louis Dumont in the key argument of *Homo hierarchicus*. I should add that, contrary to what is often implied in the orthodox (constructivist) sociology of science, asymmetry as such does not seem to me to be necessarily and universally a bad thing, or a mandatory term of pejoration in epistemology or ontology, in the field of the social sciences or elsewhere, and indeed the overwhelming cognitive (or other) merits of symmetry are not really argued for by such writers. The matter needs much further discussion and is noted in my forthcoming book, *Experimentum mundi*. Perhaps there is some underlying confusion, or at any rate an insidious identification in the minds of the symmetry-intoxicated sociologists of science between symmetry and impartiality, symmetry and objectivity, symmetry and rationality, and symmetry and equity.

33 *Tertium non datur* is a Latin phrase that means a 'third is not given', indicating that there is no third option nor alternative.

34 I did not invent this expression. The French neo-liberal economist Henri Lepage did, in his widely translated book *Demain le capitalisme* (1978).

35 American sociologist Robert K. Merton coined the term 'Matthew effect' in 1968. For further reference, see his article 'The Matthew Effect in Science: The reward and communication systems of science are considered' (1968).

36 A version of the 'Matthew effect' operated with the richer American universities in recent years when their endowments, through a mixture of successful investments and 'pull' for more donations, became not only absolutely richer, but relatively even richer compared with the less well-endowed or unendowed ones, something that seems to have completely bewitched UK university marketeers; the widening of the gap was compounded by the cuts in federal and state grants to public universities. Of course, the same mechanism applies also to the ingress of faculty or students, reputation, and so forth, where success breeds success.

37 The principle of performance-related pay (it used to be called, in the economics of wages, 'payment by results') or 'merit pay', in effect goes together with other market criteria, and may be overridden by them, within one and the same university. Thus, a full professor of English may be paid only as much as a new assistant professor of accounting in the same American university, independently of the number or measured reputation of publications, or any of the other performance criteria which, supposedly, rationally determine pay in the University of Excellence. Of course, *between* universities in the same country, wide disparities in pay for holders of the same academic rank, in similar departments, and of similar accomplishments or 'merit', have long been known in the United States, even before the postwar expansion of universities. For further reference, see the sociological study by Logan Wilson, *The Academic Man* (1942).

38 I recall a British philosopher saying many years ago, in defence of Oxbridge colleges, then as now under attack for elitism, snobbery, class bias and so on, that they represented 'the last stand of syndicalism against bureaucracy' in contemporary society, though this was not by any means the first time an academic had characterized them as 'syndicalist'. He was, or had recently been, a Trotskyite, in one of his many intellectual/political/religious/national metamorphoses, though more recently he has become widely known as a transatlantic exponent of Thomistic–Aristotelian thought, enjoying considerable international fame as a leading exponent of 'communitarianism' in moral and political philosophy. Whether they can now survive, in substance, the inroads of marketization, so zealously and relentlessly promoted by the state bureaucracy, or rather by the PM and his econocratic, pan-marketizing, e-(u)topian and generally techno-utopian advisers (being extremely partial not only to IT, but to biotech and GMOs, and now, it seems, to nanotech as well), not to mention vocal leading Oxford academics from within, remains to be seen. Everywhere else, 'the market' (unleashed by the state, backed by the state) has been mightier than syndicalism, mightier than bureaucracy, though still 'under God' according to some claims. At any rate, the dream, or project, of a Princeton-by-the-Cherwell seems a potent one, at least in one Oxford College. Perhaps they have not heard of the fears of 'Princetonization' voiced by faculty in the University of Chicago some years ago, or of 'Princeton-on-the Hudson' in a New York university a while back. I take these examples from David L. Kirp's recent book, *Shakespeare, Einstein, and the Bottom Line: The Marketing of Higher Education* (2003).

39 The UK's 'E-University', designed as a kind of electronic, online Open University for the whole world, announced by the secretary of state for education in 2000, which was expected to attract millions of students, has been closed down, with just 900 students in its books! Students in these e-university business arrangements were not only 'customers' but also 'end-users' (Is there no end to the infliction of commercial terminology on matters of education?). One may compare this grotesque failure of New Labour's E-University with the success of Old Labour's Open University, government-founded and government-funded, which it was trying to emulate and surpass but which it may never succeed in doing. It is true that less money was wasted than in the much-quoted, edifying statist horror story of the Groundnuts Scheme, but will it ever make it to neo-liberal textbooks as a comparable lesson and cautionary tale of disastrous state-driven

marketization? The lure of profit from the setting-up of what David Noble has called 'digital diploma mills' or more prestigious but still for-profit versions thereof, is perhaps too strong to preclude further attempts on similar lines being made again before too long. And the e-(u)topian vision animating so many of the government's advisers (what one might call 'cyber-chic', instead of the old-fashioned 'radical chic') will reinforce this disposition. Of course, in the 1960s and 1970s, an epoch of radical anticapitalist computopias, there was a current of thought that hoped to enlist cyber-technology in the service of a nonmarket socialism, but that was a different matter altogether. The more far-reaching e-(u)topian vision put forth with great gusto for the last three decades in the United States that online education would finish off the universities in the free market, like the vision of the need for schoolteachers and schools being eliminated by PCs and videos, has not yet come to pass and is not even within sight (though if it had come about already, it might have solved the UK university marketeers' problems at a stroke).

40 Some British universities already advertise academic vacancies in print media giving only e-mail addresses, and no postal address (or even a telephone number or a fax number) at all, as if they only existed in cyberspace, and could not be contacted or did not want to be contacted offline in the real world, as it were (or at least implying that they do not wish for any such thing, giving no physical location, while still resorting to print media, perhaps only owing to statutory requirements, for the time being). I wonder if they ever receive, or send, 'snail mail' (perhaps they dispense with it to save money and trees), though at least they did advertise vacancies offline, though perhaps reluctantly. Some academic diaries, too, list only websites for all UK universities. Perhaps they are already proleptically pretending to be wholly Virtual Universities, a higher form of academic being altogether. Unfortunately, I fear that a lot of driving to and from the campuses by faculty, administration and students still takes place regularly, at a significant ecological cost, so the virtuality is still pretty limited by hard physical and material/energetic technological reality, despite all the vaunted 'etherialization' and 'ephemeralization' that sophisticated technologies increasingly involve, with its techno-metabolism with the environment, and with the wastes and varied nefarious environmental impacts accruing therefrom.

41 The date of Caplow and McGee's publication (1958) shows how far back some of the more farcical present practices, or alleged practices – and indeed we could truly say *si non è vero, è ben trovato* – of academic evaluation go.

42 Edinburgh University was ranked 53 in the list of world universities previously referred to.

43 The sociologist Arnold Gehlen characterized the person in modern society as an 'institution in one case' in his book *Man in the Age of Technology* (1980). Today, the individual academic is becoming (induced to become by multitudinous pressures not least from the university powers-that-be, as well as by virtue of their belonging to the winner-takes-all market society) not just a shareholder in a firm, but a firm in one person, and this seems to be a general trend in contemporary society. This may signify, in due course, perhaps another shift in the long-lived 'category of the person' (Mauss 1985), more particularly from the autonomous citizen under the Moral Law to the person-as-firm, or the market-in-one-person-case. There are now learned neo-liberal works on the 'economics of mind' (this was before the emergence of experimental 'neuro-economics', linking neurons in the brain-market to purchases in the outer market), on the human mind or the human brain as market-like or economorphic (so to some extent human psychology could be seen as a branch of micro-economics, or at least as structurally isomorphic to it), which could be glossed too as addressing the economics of the person, the market within the brain/mind, acting in the market or markets without (commodity markets, political markets, knowledge markets, religious salvation markets, genetic (soon gene chip)/genomic/proteonomic services and capital markets – to refer only to markets on which there is a considerable, influential academic literature, social scientific or philosophical).

Is there room in the conceptual map of the marketizing elites for an essentially nonmarket core of a person, as for an essentially nonmarket domain of the wider society? In the vision of Downing Street advisers, as for some ultra-liberals, perhaps not.

44 This phenomenon was extensively discussed in the important study by Christopher Jencks and David Riesman (two Harvard academics) *The Academic Revolution* (1968). There is of course much else relevant to a comprehensive study of the American research university at the end of the golden age in this work.

45 *Raison d'être* is a French phrase that means 'reason for being'.

46 B. Charlton, in a recent essay on the condition of British universities, has argued that, in effect, the universities have been redesigned to supply *middle managers* to the nation, or rather to the 'workforce', which of course sounds much less attractive than the chancellor's vision, but is surely closer to the truth. (I am grateful to Grahame Lock, of Nijmegen University in the Netherlands, for drawing my attention to this essay.) The increase of the studentry to 40–50 per cent of the relevant age cohorts would roughly match the proportion of the workforce that may be characterized as performing managerial functions of one sort or another in the 'real' economy, at this time or at any rate in the next decade or so. There was a time when promotional brochures of UK universities spoke of the training for 'leadership' students would receive there (the assumption being that there were sufficient leadership positions to be filled in the wider society). The term 'leadership' has fallen into comparative disuse in general parlance, but today, clearly, students are being trained for what one might call in parallel fashion, '*managership*', to meet the needs of an increasingly managerialized economy, though this is not official, and 'entrepreneurship', the avowed ideal, sounds better than 'managership'. To be sure, as 'flexibility' increases, the greater the numbers of the downsized, the more people have to try for entrepreneurship. And clearly the universities are expected to help produce or reproduce the 'cybertariat'. The word was coined, though not really defined explicitly, by Ursula Huws in a paper substantially reproduced in her book *The Making of a Cybertariat: Virtual Work in a Real World* (2003), though special incentives or pressures for their production are hardly needed in this technological epoch. Sufficient numbers of cybertarians may well spring up, without official bidding, almost from infancy if not from the womb, as some would have it, in an almost literally cyber-reared, indeed soon to be cyber-gestated world (through electronically mediated training of the foetus and eventually the emergence of the full-fledged 'cyborg child'). In fact, when 40 per cent or more of the workforce fulfil managerial roles, clearly a good proportion of them could be classed as a kind of *manageriat(e),* rather than a 'cyberbourgeoisie'. But in addition, the universities must provide a stratum of research scientists, doctoral students and postdoctoral research fellows (these will have to be far more strenuously nurtured). The latter, as in America, in the hard sciences at least, may come increasingly from abroad (outsourced inwards, so to speak), rather than from native 'human resources', until such a time as these foreign sources of trained talent dry up in turn, though ever more sophisticated Roboscientists or Science Robots, as they have been called in a lengthy article in *Nature* by a team of scientists describing the cognitive successes of the first effective members of this new taxon in sociocultural, and in epistemic, ontology (see *Addendum 10*), which may be increasingly drawn upon to stave off that day, replacing the work done by vanishing graduate students and postdocs, brainware by software, manual labour by robotic arms and feet, and their combinations. Even in virtual reality, humans will soon be able to feel or at least feel that they feel, touch and grasp real-world objects in a 3-D world. Whatever may be said about the idea of a nation of entrepreneurs, it could also be said that in every Western country there are emerging nations of proletarians, in the sense of the term enunciated by the German Catholic sociologist Goetz Briefs decades ago, when he defined the proletariat as the body of persons wholly or largely dependent for their livelihoods on shifting and insecure labour markets. For one sure thing about labour markets (as well as other types of markets) under turbo-capitalism, or techno-capitalism, is their shifting and insecure character, glossed very nicely as 'flexibility', 'innovation', 'creativity', 'creative destruction', 'accelerated change', and the like, to be embraced with enthusiasm, with the end of job security and careers for life, the shrinking of the public sector, and other concurrent transformations (the instability of financial markets is another matter).

47 'Entrepreneur' has become, in the last decade or two, perhaps the supreme honorific term of the age, and all too widely applied as a 'hurrah word', not only in general parlance, but even within social science discourse. Schumpeter's famous eulogy of the entrepreneur (first published in 1912) was partly inspired by the Carlylean vision of 'heroes', as shapers of history, and something of this Carlylean heroization persists in the current valuations, though the 'heroes of consumption' (pop stars, the top range of entertainers of all kinds, celebrities, the famous for being famous) also enjoy high status and wealth, though they may be regarded as entrepreneurs of sorts. Leo Lowenthal's famous succession model of cultural change, involving a major shift from the predominance of 'heroes of production' to the predominance of 'heroes of consumption', has thus been partially invalidated by the course of social transformation under the last three governments or so. To be sure, the phrase 'heroes of production' should be taken in a wide sense, for many of them could perhaps be seen more as 'heroes of business', 'heroes of management', 'heroes of finance', 'heroes of asset-stripping', 'heroes of organization reengineering', 'heroes of creative accounting', 'heroes of "paper entrepreneurialism"' (to borrow the American economist Robert Reich's expression), 'heroes of crony capitalism', or 'heroes of predatory capitalism' than 'heroes of production' in any strict sense. Of course, in the e-economy there are a number of outstanding inventor-entrepreneurs, and the three-in-one scientist-engineer-entrepreneur composite is a salient feature of the New Economy and bio-economy landscape of our time. One should bear in mind that the Schumpeterian entrepreneur was not exclusively or even primarily associated with the deployment of new technology. Both in his first formulation of the concept of entrepreneur and in his late work, entrepreneurship could also involve other domains such as the discovery of new resources, the exploitation of new markets, or the advantages of new organizational forms. Of course, the celebrity academic superstar is both a hero of knowledge-production and a hero of consumption in the entertainment field.

Even doing good in the public realm outside the state is not characterized any longer as doing good, or altruism, or philanthropy, charitable work, or the exercise of benevolence, or Christian action, or the exercise of Christian charity or *agape*, or the exercise of beneficence, or saintly conduct, or paying a debt to others, or the practice of fraternity, or as demonstrating solidarity with the less fortunate, or as the love of humanity in the Enlightenment and Positivist tradition, but just as 'social entrepreneurship'! The good are not any longer to be seen as 'creative altruists', as Sorokin called them in his sociological studies of ordinary American philanthropists in the 1950s. Not surprisingly, they were unfashionable then, totally unfashionable now in every ideological quadrant, within or without sociology, let alone economics, it surely being the case that neither discipline barely deigns to recognize their existence, to put it mildly. No, 'altruism' is definitely suspect, conceptually and lexically, throughout the social and human sciences, in this way a true mirror of the operational code of the market. More than suspect, to be exorcized, as in the formulation 'reciprocal altruism' in ultra-Darwinian sociobiology, which is not at all what was meant by altruism before, so that putative altruists turn out to be, reassuringly, just another species, another subset, of entrepreneurs, specifically 'social entrepreneurs', like it or not! This astonishing and imperious stretching of the category 'entrepreneur', where not so long ago we had an ample and resonant vocabulary of religious and moral praise with Christian or Enlightenment roots, at our disposal to refer to good voluntary action towards the unfortunate or deprived, which moral vocabulary had served us for a long time, indeed for centuries, demonstrates how the enforcement of market discipline operates even in our hallowed moral idioms, which must yield to the usurpation of prepotent market categories. They thus appear increasingly, if not as the sole legitimate ones, at any rate as enjoying certain power of privileged illumination and ratification. Why is it that no one refers to 'anti-social entrepreneurs', entrepreneurs of evil, entrepreneurs of media malice or 'feeding frenzy', entrepreneurs of organized crime, media entrepreneurs of xenophobia, and the like? There was a time when a saying such as '*on n'aime rien tant que ceux qu'on a vu humiliés*' could be understood without any reference to the categories of the market (Durkheim characterized socialism, as movement or doctrine as stemming from a '*cri de douleur*').

Perhaps it would be better to go along with the tide and call the Welfare State the '*social entrepreneurial state*' in partnership with 'social entrepreneurs' at large, in 'civil society'. With luck, more money might be forthcoming, and the sense of unease that seems associated with it, mitigated. Anyway, this proposal for rebaptizing the Welfare State, this little piece of 'linguistic entrepreneurship', is offered on 'open access', with some trepidation, for it would be better if we at least tried to halt if not reverse the trend towards the marketization of language. Difficult, to be sure, for market orthodoxy, like all other orthodoxies, is accompanied, and partly enforced by, a kind of orthoglossia. But maybe I am complaining too much, for a cyber-executive and cyber-millionaire recently proclaimed that money 'is the root of all good', explicitly subscribing to the Ayn Rand ethical doctrine of Absolute Selfishness (a doctrine that informs more than a philosophical school, a kind of American cult, among some professionals and entrepreneurs, furnishing a most satisfying 'theodicy of good fortune', to add to the more classical versions). Of course, quite a number of digerati incline towards a Nozickian minimal state, or a version of anarcho-capitalism, or cyber-libertarianism, or 'crypto-anarchy'. Some say this is inevitable, given the nature of the technologies involved, though the United States is most concerned to prevent this as well as being the home of those most committed to fostering it, so the state will have to be shrunk much further, to a degree not yet reached anywhere in an industrial society, even in the Information Age. As yet, few have seriously argued in the media for the privatization of the armed forces and the entirety of the police, as the anarcho-capitalist economist-ideologue David Friedman, son of Milton Friedman, did so forcefully.

Against this background, as a beneficiary of 'rents of ability' (as the early Fabians liked to say), as a 'rent-seeker' (all exiles are, I suppose, in a way, *ceteris paribus*), as a 'rentier', in the denigratory parlance of the times (though Pareto had already sharply contrasted 'entrepreneurs' and 'rentiers'), I take some comfort from the staunchly held view of Ludwig von Mises, one of the greatest of 'Austrians', still a major influence in an important strand of contemporary economic-liberal thought, that everyone, absolutely everyone, is, willy-nilly, an entrepreneur, a risk-taking, or, better, uncertainty-facing agent (uncertainty being the primary category in this perspective), necessarily and universally so, by virtue of the mere fact of being an acting person in the world.

48 In a way, the RAE, carried out yearly for quite a number of years in the UK, could be seen as, de facto, enforcing in everything but the name, a partial, sectoral, *labour code for academics*, treating academia, without prior warning, precedent or warrant, as an instance of a *continuous-process industry*, as it were, where maturing reflection, time to learn, time to master new languages, for example, or even taking care to avoid adding to the existing load of intellectual pollution, to contribute to the conversation of academic and intellectual life, and the like, were disqualified and tacitly negatively valued, as they should be, perhaps, in industries that are effectively continuous-process industries, though not all industries could be properly so called as yet, though undoubtedly the trend is going that way, with the assistance of genetic engineering and biotechnology in general, and directly or indirectly with the computationalization of all operations. The redefining of contractual conditions has added further downgrading to the informal Academic Labour Code in still physically existing university-factories (even outside the domain of the 'digital diploma mills' proper), that came into being particularly via the RAE, the code of university-based knowledge-industry labour, on closely parallel lines to the recent market-friendly, labour-unfriendly, flexibility-maximizing, labour codes of 'real' industry in general. For some analysts though, an ever-greater proportion of industries deserve to fall under the rubric of 'knowledge-industry', too. Universities, and certainly research universities, are still a somewhat peculiar species of the knowledge industry, but their peculiarities may have outlived their usefulness, until they become wholly and exclusively market-driven research universities with state conscription into the industrial, postindustrial or hyperindustrial world's

ever more inclusive, pervasive and inescapable 'treadmill of production' (a concept originally formulated by the sociologist Allan Schnaiberg in his 1980 book *The Environment: From Surplus to Scarcity*), in short, from which one no academic inmate must be allowed to get off. To be sure, no one is forcing them to be academics; if they do not like the treadmill of academic knowledge production, they can relocate and redeploy whatever saleable 'skills', if any, they may possess to other branches of the society-wide or world-wide technological treadmill of production, of the World Technomarket, if they want to survive, so there's a 'choice'!

Given the dire consequences of the operation, which must have been the whole point of it, it seems absurd, indeed, perverse, that it was called, is still called, an 'exercise'. Live ammunition, as it were, or at least real, substantial, financial sanctions, were deployed, though words alone *can* kill and most certainly wound, departments were injured, individuals distressed, morale undermined, opportunities denied, and intellectual life further degraded. More, it was a scientific Grand Guignol performance, a technocratic version of the Visitations of the Inquisition, an academic enactment of a kind of *Modern Times*, than something fitting for the House of Intellect. But then, academia is seen from Whitehall more in factory and business terms than anything else: from the Academia of Scholars (in its ideals at least) to the Turbo-Capitalist Academia of Entrepreneurs (capitalize on your knowledge!) in one generation! Another imperative of universities today is of course the internalization of the entrepreneurial ethic, with knowledge as the commodity/subject by the faculty. After all, in a nation of entrepreneurs, the academics must enlist, too. In this climate, it is not surprising that some academics, taking their lesson to heart, turn into many-sided entrepreneurs, not infrequently often at the expense of the universities themselves, though generally universities appear quite indulgent towards them (certainly compared to the sanctions inflicted on harmless, 'unproductive' addicts of the music of ideas, or 'unproductive', devoted and effective teachers).

Now more than half of the university departments of chemistry (not sociology, or media studies, or cultural studies) in the UK are threatened with closure because they did not get the top grades in the last RAE 'exercise'. It seems that if you do not get the top grades in this kind of operation, you have failed, and departmental closure may be not far off, in line with the rapid 'elimination of the unfit' in Darwinian academic capitalism: why not just award two grades, Pass (top grades), and Fail (anything below a '5', starred or unstarred)? This turns academia essentially into a game of winning or losing, like a competitive sport, nothing else, most being losers through a kind of *Law of Almost Universal Failure* of university departments. Here as in the winner-takes-all economy as a whole, but through political agency ultimately, *le mieux est l'ennemi du bien*. Of course, as far as students are concerned, precisely the reverse is true, for the tendency is to give everybody prizes. In most universities, even in the 'best' universities, as the president of Harvard University, where presumably most or all departments would rate the equivalent of the RAE's 5 or 5*, complained not so long ago, something like a Law of Grade Inflation or Law of Prizes for Almost Everyone, applies to students, whereby an ever-greater proportion of graduates get awarded the upper grades; maybe there is an objective basis for this, such as the 'Flynn effect', named after the psychologist James Flynn, who first reported on the convergence of independent findings over two or three decades of rising IQ levels in young age cohorts in advanced Western societies, which has generated further research to test these results, which does appear to have further confirmed the overall appraisal.

Perhaps some sectors of the universities, at least, could apply to an appropriate regulatory body for reclassification as branches of some *slow process industry* or other. There may well be still some left under something like that description, instead of the continuous-process industry model. And, better still, even join the Slow movements that have sprung up in the last decade or so, starting in Italy – 'slow food', 'slow cities', but also now elsewhere 'slow art' [. . .] 'idle curiosity' (Veblen). Instead of the unhappy 'culture wars', or the 'science wars' of recent years, join the 'time wars' (more latent than the book of Jeremy Rifkin (1987) with the same title would have led one to expect), the Contemporary Simplicity Movement (in the United States), or, better, the Society for the Deceleration of Time.

49 Note that, at the local level, universities, or university managements, have often been, and continue to be, despoilers of their environment, urban or rural, from the ecological, aesthetic and social points of view, against the protests of citizens, students, and faculty. To be sure, many European cities seem as eager for universities as for international airports, regardless of the known manifold environmental detriments of both. I am not sure whether marketized universities also clamour for airports, the more so as they become more corporatized and, like corporations everywhere ever more delocalized, but in any case, increasingly wedded, like corporations (almost) everywhere, and governments (supposedly in response to consumer demand), to hyperproductivity and hypermobility, they are coming to look and feel like airports, with ever more transient populations of students and faculty (turbo-profs, stellar faculty always on the move, casualized academic labour, holders of insecure academic Mcjobs, worried about their next equally insecure academic or para-academic job), increasingly fittingly describable as '*non-lieux*', as 'non-places', in the well-known terminology of the French social anthropologist Marc Augé. To be sure, if it were true that in the information society social space becomes overwhelmingly a 'space of flows' (Castells 1989), presumably every former 'place' within it tends to become a kind of 'was-once-a-place', now a 'non-place', in this sense, no more than a conduit of flows, bereft of intrinsic, non-fluxional, binding qualities. On the concept of 'hypermobility', see the article 'Hypermobility' (2000) by geographer John Adams, as well as his Organization for Economic Cooperation and Development report, *The Social Implications of Hypermobility* (1999).

REFERENCES

Ackerman, Diane. 1990. *A Natural History of the Senses*. New York: Vintage Books.

Adams, John. 2000. 'Hypermobility'. *Prospect*, 20 March. https://www.prospectmagazine.co.uk/magazine/hypermobility.

———. 1999. *The Social Implications of Hypermobility: Speculations about the Social Consequences of the OECD Scenarios for Environmentally Sustainable Transport and Business-as-Usual Trend Projections*, 75–113. Paris: OECD.

Agar, Nicholas. 2004. *Liberal Eugenics: In Defence of Human Enhancement*. Malden: Blackwell.

———. 2002. *Perfect Copy: Unravelling the Cloning Debate*. Cambridge: Icon Books.

Alexander, Brian. 2003. *The Rapture: How Biotech became the New Religion*. New York: Basic Books.

Almeling, Rene. 2011. *Sex Cells: The Medical Market for Eggs and Sperm*. Berkeley: University of California Press.

Anarchism and Transhumanism. 2015. 'Mission Statement'. http://anarcho-transhumanism.net/about/.

Angell, Marcia. 2004a. *The Truth about the Drug Companies: How They Deceive Us and What to Do about It*. New York: Random House.

———. 2004b. 'The Truth about the Drug Companies'. *New York Review of Books*, 15 July. http://www.nybooks.com/articles/2004/07/15/the-truth-about-the-drug-companies/.

Annan, Noel. 1955. 'The Intellectual Aristocracy'. In *Studies in Social History: A Tribute to G. M. Trevelyan*, edited by J. H. Plumb, 243–87. London: Longmans.

Annas, Julia. 2011. *Intelligent Virtue*. Oxford: Oxford University Press.

Arestis, Philip, ed. 2001. *A Biographical Dictionary of Dissenting Economists*. Revised edition. Cheltenham: Elgar.

Atlan, Henri. 2005. *L'utérus artificial*. Paris: Le Seuil.

Axe, David. 2011. 'One in 50 Troops in Afghanistan is a Robot'. *Wired*, 7 February. https://www.wired.com/2011/02/1-in-50-troops-robots/. Accessed 19 July 2018.

Ayres, Robert U. and Jörg Walter. 1991. 'The Greenhouse Effect: Damages, Costs and Abatement'. *Environmental and Resource Economics* 1 (3): 237–70.

Bailey, Ronald. 2005. *Liberation Biology: The Scientific and Moral Case for the Biotech Revolution*. New York: Prometheus Books.

Bainbridge, William Sims. 2007. *Nanoconvergence: The Unity of Nanoscience, Biotechnology, Information Technology, and Cognitive Science*. Upper Saddle River, NJ: Pearson Education.

Baker, Robin. 1999. *Sex in the Future: Ancient Urges Meet Future Technology*. London: Macmillan.

Baldamus, W. 1966. *The Category of Pragmatic Knowledge in Sociological Analysis*. Birmingham: University of Birmingham.

Barton, Nicholas FRCS. 2013. 'Letter'. *Times*, 7 February.

Becker, Lawrence C. 1998. *A New Stoicism*. Princeton, NJ: Princeton University Press.

Bell, Daniel. 1976. *The Cultural Contradictions of Capitalism*. New York: Basic Books.

———. 1973. *The Coming of Post-Industrial Society: A Venture in Social Forecasting*. New York: Basic Books.

———. 1970. 'The Cultural Contradictions of Capitalism'. *Public Interest* 21: 16–43.

Bentham, Jeremy. 1843. *The Works of Jeremy Bentham*. Vol. 4. London: Simpkin, Marshall.

Berlin, Isaiah. 1997 [1959]. *The Crooked Timber of Humanity*. Princeton, NJ: Princeton University Press.

Bernal, John Desmond. 1939. *The Social Function of Science*. London: Routledge.

———.1929. *The World, the Flesh and the Devil: An Inquiry into the Future of the Three Enemies of the Rational Soul.* London: Kegan Paul.

Bertell, Rosalie. 2000. *Planet Earth: The Latest Weapon of War.* London: Quartet Books.

Bluestone, Natalie Harris. 1987. *Women and the Ideal Society: Plato's Republic and Modern Myths of Gender.* Oxford: Berg.

Bok, Derek. 2003. *Universities in the Marketplace: The Commercialization of Higher Education.* Princeton, NJ: Princeton University Press.

Bouglé, Célestin. 1899. *Les idées égalitaires: étude sociologique.* Paris: Felix Alcan.

Boulding, Kenneth E. 1964. *The Meaning of the Twentieth Century.* New York: Harper & Row.

Bowley, Graham. 2011. 'The New Speed of Money, Reshaping Markets'. *New York Times*, 1 January. http://www.nytimes.com/2011/01/02/business/02speed.html.

Breton, Philippe. 1990. *La tribu informatique.* Paris: Métailié.

Broad, C.D. 1925. *The Mind and its Place in Nature.* New York: Harcourt, Brace.

Broderick, Damien. 1999. *The Last Mortal Generation: How Science Will Alter Our Lives in the 21st Century.* Sydney: New Holland.

Brown, Andrew. 1999. *The Darwin Wars: The Scientific Battle for the Soul of Man.* London: Simon & Schuster.

Brown, Harold I. 1990. 'Prospective Realism'. *Studies in History and Philosophy of Science* 21 (2): 211–42.

Bruckner, Pascal. 2000. *L'Euphorie perpétuelle: Essai sur le devoir de bonheur [Perpetual Euphoria: On the Duty to be Happy].* Paris: Grasset & Fasquelle.

Burke, Kenneth. 1954 [1935]. *Permanence and Change: An Anatomy of Purpose.* Indianapolis: Bobbs-Merrill.

Calvin, William H. 2008. *Global Fever: How to Treat Climate Change.* Chicago, IL: University of Chicago Press.

———. 2004. *A Brief History of the Mind: From Apes to Intellect and Beyond.* Oxford: Oxford University Press.

Campbell, John H. 1995. 'The Moral Imperative of Our Future Evolution'. In *Evolution and Human Values*, edited by Robert Wesson and Patricia A. Williams, 79–114. Amsterdam: Rodopi.

Čapek, Milič. 1961. *The Philosophical Impact of Contemporary Physics.* New York: Van Nostrand.

Caplow, Theodore and Reece J. McGee. 1958. *The Academic Marketplace.* New York: Basic Books.

Carlson, Elof Axel. 1981. *Genes, Radiation and Society: The Life and Work of H. J. Muller.* Ithaca: Cornell University Press.

Carnap, Rudolf. 1936. 'Testability and Meaning'. *Philosophy of Science* 3 (4): 419–71.

Carney, Scott. 2011. *The Red Market: On the Trail of the World's Organ Brokers, Bone Thieves, Blood Farmers, and Child Traffickers.* New York: William Morrow.

Carpenter, Edmund. 1974 [1973]. *Oh, What a Blow that Phantom Gave Me!* New York: Holt, Rinehart and Winston.

Cartwright, Jon. 2010. 'Rise of the Robots and the Future of War'. *The Guardian*, 20 November. https://www.theguardian.com/technology/2010/nov/21/military-robots-autonomous-machines. Accessed 18 July 2018.

Castells, Manuel. 1989. *The Informational City: Information Technology, Economic Restructuring, and the Urban-Regional Process.* Oxford: B. Blackwell.

Chang, Ruth, ed. 1998. *Incommensurability, Incomparability, and Practical Reason.* Cambridge, MA: Harvard University Press.

Clynes, Manfred. 1977. *Sentics: The Touch of Emotions.* New York: Doubleday/Anchor Press.

Cook, Michael. 2011. 'Are We Morally Obliged to Participate in Research?' *BioEdge*, 22 October. https://www.bioedge.org/bioethics/are_we_morally_obliged_to_participate_in_research/9793. Accessed 18 July 2018.

Collier, Andrew. 1979. 'In Defence of Epistemology'. In *Issues in Marxist Philosophy*. Vol. 3, edited by John Mepham and David Hillel Reuben, 55–106. Brighton: Harvester Press.

Collingwood, Robin George. 1940. *An Essay on Metaphysics.* Oxford: Clarendon Press.

Corbin, Alain. 1986. *The Foul and the Fragrant: Odor and the French Social Imagination.* Cambridge, MA: Harvard University Press.

Cornford, F. M. 1907. *Thucydides Mythistoricus.* London: Edward Arnold.
Craig, James C. and Gary B. Rollman. 1999. 'Somesthesis'. *Annual Review of Psychology* 50: 305–31.
Cushing, Ellen. 2013. 'Amazon Mechanical Turk: The Digital Sweatshop'. *Utne Reader*, January/February. http://www.utne.com/science-and-technology/amazon-mechanical-turk-zm0z13jfzlin.
Daily Telegraph. 2007. December 11.
Danto, Arthur. 1999. *The Body/Body Problem: Selected Essays.* Berkeley: University of California Press.
Darlington, Cyril Dean. 1969. *The Evolution of Man and Society.* London: George Allen & Unwin.
Darwin, Leonard. 1926. *The Need for Eugenic Reform.* London: John Murray.
Dasgupta, Partha, Karl-Göran Mäler and A. Vercelli, eds. 1997. *The Economics of Transnational Commons.* Oxford: Clarendon Press.
Davis-Floyd, Robbie and Joseph Dumit, eds. 1998. *Cyborg Babies: From Techno-Sex to Techno-Tots.* New York: Routledge.
de Garis, Hugo. 2005. *The Artilect War: Cosmists vs. Terrans: A Bitter Controversy Concerning Whether Humanity Should Build Godlike Massively Intelligent Machines.* Palm Springs: ETC.
Dijksterhuis, Eduard Jan. 1961. *The Mechanization of the World Picture: Pythagoras to Newton.* New York: Oxford University Press.
Dowd, Douglas. 1989. *The Waste of Nations: Dysfunction in the World Economy.* Boulder, CO: Westview Press.
Dubos, René Jules. 1959a. 'Medical Utopias'. *Current Work and Controversies* 88 (3): 410–24.
———. 1959b. *Mirage of Health: Utopias, Progress, and Biological Change.* New York: Harper & Brothers.
Dumont, Louis. 1977. *Homo aequalis: genèse et épanouissement de l'idéologie économique.* Paris: Gallimard.
Dupré, John. 1993. *The Disorder of Things: Metaphysical Foundations of the Disunity of Science.* Cambridge, MA: Harvard University Press.
Durkheim, Émile. 1912. *Les formes élémentaires de la vie religieuse: Le system totémique en Australie.* Paris: Felix Alcan.
Duster, Troy. 1990. *Backdoor to Eugenics.* New York: Routledge.
Dworkin, Ronald. 1993. *Life's Dominion: An Argument about Abortion, Euthanasia, and Individual Freedom.* New York: Knopf.
Dyens, Ollivier. 2001. *Metal and Flesh: The Evolution of Man, Technology Takes Over.* Cambridge, MA: MIT Press.
Eichner, Alfred S., ed. 1983. *Why Economics is Not Yet a Science.* New York: M. E. Sharpe.
Ellul, Jacques. 1954. *La Technique ou l'Enjeu du siècle.* Paris: Armand Colin.
Fackenheim, Emil L. 1961. *Metaphysics and Historicity, The Aquinas Lecture.* Milwaukee: Marquette University Press.
Fairclough, Norman. 2000. *New Labour, New Language?* London: Routledge.
Fassler, Joe. 2012. 'Can the Computers at Narrative Science Replace Paid Writers?' *Atlantic*, 12 April. https://www.theatlantic.com/entertainment/archive/2012/04/can-the-computers-at-narrative-science-replace-paid-writers/255631/.
Ferkiss, Victor. 1980. 'Technology and Culture: Gnosticism, Naturalism, and Incarnational Integration'. *CrossCurrents* 30 (1): 13–26.
Ferrand, Michèle. 2013. 'La pilule est à l'origine de certains risques, mais elle est utile'. *Le Monde*, 2 July. http://www.lemonde.fr/culture/article/2013/02/07/la-pilule-est-a-l-origine-de-certains-risques-mais-elle-est-utile_1828675_3246.html.
Financial Times. 2004. January 15.
Fletcher, Joseph. 1974. *The Ethics of Genetic Control: Ending Reproductive Roulette.* New York: Anchor Books.
Flynn, James R. 2007. *What Is Intelligence? Beyond the Flynn Effect.* Cambridge: Cambridge University Press.
Foucault, Michel. 1977. *Discipline and Punish: The Birth of the Prison.* New York: Pantheon Books.
Foucault, Michel and Immanuel Kant. 2008. *Kant: Anthropologie du point de vue pragmatique & Foucault: Introduction à l'Anthropologie*, edited by D. Defert, Fr. Ewald and F. Gros. Paris: J. Vrin.
Frank, Robert H. and Philip J. Cook. 1995. *The Winner-Take-All Society: Why the Few at the Top Get So Much More Than the Rest of Us.* New York: The Free Press.

Franzen, Jonathan. 2011. 'Liking Is for Cowards. Go for What Hurts'. *New York Times*, May 28. http://www.nytimes.com/2011/05/29/opinion/29franzen.html.

Fraassen, Bas C. van. 2002. *The Empirical Stance*. New Haven, CT: Yale University Press.

———. 1980. *The Scientific Image*. Oxford: Clarendon Press.

'French Court Rules against Frozen Couple'. 2002. *BBC News*, 13 March. http://news.bbc.co.uk/2/hi/europe/1870301.stm. Accessed 19 July 2018.

Galison, Peter and David J. Stump, eds. 1996. *The Disunity of Science: Boundaries, Contexts, and Power*. Stanford: Stanford University Press.

Galton, David. 2002. *Eugenics: The Future of Human Life in the 21st Century*. London: Abacus.

Galton, Francis. 1869. *Hereditary Genius: An Inquiry into its Laws and Consequences*. London: Macmillan.

Gavaghan, Colin. 2007. *Defending the Genetic Supermarket: Law and Ethics of Selecting the Next Generation*. Abingdon: Routledge-Cavendish.

Geary, James. 2002. *The Body Electric: An Anatomy of the New Bionic Senses*. New Brunswick, NJ: Rutgers University Press.

Gehlen, Arnold. 1980. *Man in the Age of Technology*. New York: Columbia University Press.

Geiger, Theodore. 1969. *Theodore Geiger On Social Order and Mass Society*, edited by Renate Mayntz. Chicago, IL: University of Chicago Press.

George, Robert P. and Christopher Tollefsen. 2008. *Embryo: A Defense of Human Life*. New York: Doubleday.

Gibbons, Michael, et al. 1994. *The New Production of Knowledge: The Dynamics of Science and Research in Contemporary Societies*. London: Sage.

Gibson, James J. 1966. *The Senses Considered as Perceptual Systems*. Boston, MA: Houghton Mifflin.

Gilbert, Daniel. 2006. *Stumbling on Happiness*. New York: Knopf.

Ginsberg, Morris. 1934. *Sociology*. London: Thornton Butterworth.

Gleason, Bill. 2011. 'Do People Have a Moral Obligation to Be Research Subjects?' *Chronicle of Higher Education*, 14 October. http://www.chronicle.com/blogs/brainstorm/do-people-have-a-moral-obligation-to-participate-in-research/40251.

Good, I. J. 1965. 'Speculations Concerning the First Ultraintelligent Machine'. *Advances in Computers* 6: 31–88.

Good, I. J., 1962. Alan James Mayne and John Maynard Smith, eds. *The Scientist Speculates: An Anthology of Partly Baked Ideas*. London: Heinemann.

Goody, Jack. 1983. *The Development of the Family and Marriage in Europe*. Cambridge: Cambridge University Press.

Goetz, Thomas. 2012. 'Why Doctors Need to Embrace Their Digital Future Now'. *Wired Magazine*, 31 January. https://www.wired.com/2012/01/st_topolqa/. Accessed 19 July 2018.

Gordon Walker, Patrick. 1951. *Restatement of Liberty*. London: Hutchinson.

Gould, Eric. 2003. *The University in a Corporate Culture*. New Haven, CT: Yale University Press.

Gould, Kenneth A., David N. Pellow and Allan Schnaiberg. 2008. *The Treadmill of Production: Injustice and Unsustainability in the Global Economy*. Boulder, CO: Paradigm.

Grant, Ruth W. 2012. *Strings Attached: Untangling the Ethics of Incentives*. Princeton, NJ: Princeton University Press.

Gray, John. 1963. *I & II Kings: A Commentary*. Philadelphia: Westminster Press.

Grayling, A. J. 2004. 'The Reason of Things'. *Times*, 31 January.

Greenfield, Susan. 2003. *Tomorrow's People: How 21st Century Technology Is Changing the Way We Think and Feel*. London: Allen Lane.

Hakim, Catherine. 2012. *Honey Money: Why Attractiveness Is the Key to Success*. London: Penguin.

Haldane, Charlotte. 1926. *Man's World*. London: Chatto & Windus.

Haldane, J. B. S. 1924. *Daedalus; or, Science and the Future*. New York: E. P. Dutton.

Hamilton, W. D. 2002. *Narrow Roads of Gene Land: Evolution of Sex*. Vol. 2. Oxford: Oxford University Press.

Hamermesh, Daniel S. 2013. *Beauty Pays: Why Attractive People Are More Successful*. Princeton, NJ: Princeton University Press.

Hare, William Loftus. 1923. *Mysticism of the East and West: Studies in Mystical and Moral Philosophy*. London: Jonathan Cape.

Harris, Marvin. 1964. *The Nature of Cultural Things*. New York: Random House.

Hartley, L. P. 1960. *Facial Justice*. London: Hamish Hamilton.

Hawking, Stephen W. and Leonard Mlodinow. 2010. *The Grand Design*. New York: Bantam Books.

Healy, David. 1997. *The Anti-Depressant Era*. Cambridge, MA: Harvard University Press.

Hellebust, Rolf. 2003. *Flesh to Metal: Soviet Literature and the Alchemy of Revolution*. Ithaca: Cornell University Press.

Heylighen, Francis. 2007. 'The Global Superorganism: An Evolutionary-Cybernetic Model of the Emerging Network Society'. *Social Evolution and History* 6 (1): 57–117.

Hirschman, Albert O. 1970. *Exit, Voice, and Loyalty: Responses to Declines in Firms, Organizations, and States*. Cambridge, MA: Harvard University Press.

Hobbes, Thomas. 1651. *Leviathan*. London: Andrew Crooke.

Hoyle, Fred. 1964. *Of Men and Galaxies*. Seattle: University of Washington Press.

Hurley, James. 2011. 'Employers Call for 'Enterprise Curriculum'. *Telegraph*, 26 July. http://www.telegraph.co.uk/finance/yourbusiness/8661915/Employers-call-for-enterprise-curriculum.html. Accessed 19 July 2018.

Huws, Ursula. 2003. *The Making of a Cybertariat: Virtual Work in a Real World*. London: The Merlin Press.

Huxley, Aldous. 1932. *Brave New World*. London: Chatto & Windus.

Huxley, Julian. 1970. *Memories*. London: George Allen & Unwin.

Inglehart, Ronald. 1977. *The Silent Revolution: Changing Values and Political Styles among Western Publics*. Princeton, NJ: Princeton University Press.

Jay, Martin. 1993. *Downcast Eyes: The Denigration of Vision in Twentieth-Century French Thought*. Berkeley: University of California Press.

———. 1984. *Marxism and Totality: The Adventures of a Concept from Lukács to Habermas*. Berkeley: University of California Press.

Jencks, Christopher and David Riesman. 1968. *The Academic Revolution*. New York: Doubleday.

Jenkins, Simon. 2004. *Times*. January.

Johnson, Benjamin, Patrick Kavanagh and Kevin Mattson. 2003. *Steal This University: The Rise of the Corporate University and the Academic Labor Movement*. New York: Routledge.

Jonas, Hans. 1966. 'The Nobility of Sight: A Study in the Phenomenology of the Senses'. In *The Phenomenon of Life: Toward a Philosophical Biology*, 135–52. New York: Harper and Row.

Jones, Gareth. 2001. *Clones: The Clowns of Technology?* Carlisle: Paternoster Press.

Jones, Steven. 2003. *Y: The Descent of Men*. Boston, MA: Houghton Mifflin.

Kass, Leon R. 1998a. 'The Wisdom of Repugnance'. In *The Ethics of Human Cloning*, edited by Leon Kass and James Q. Wilson, 3–59. Washington DC: AEI Press.

———. 1998b. 'The Wisdom of Repugnance'. In *The Human Cloning Debate*. Second edition, edited by Glenn McGee, 68–106. Berkeley: Berkeley Hill Books.

———. 1997. 'The Wisdom of Repugnance: Why We Should Ban the Cloning of Humans'. *The New Republic* 216 (22): 17–26.

Kaufmann, Felix. 1944. *Methodology of the Social Sciences*. London: Oxford University Press.

Kearney, Richard. 1994. *Continental Philosophy in the 20th Century*. Vol. 8. London: Routledge.

Kelly, Kevin. 1994. *Out of Control: The Rise of Neo-Biological Civilization*. Boston, MA: Addison-Wesley.

Kirp, David L. 2003. *Shakespeare, Einstein, and the Bottom Line: The Marketing of Higher Education*. Cambridge, MA: Harvard University Press.

Kluckhohn, Clyde. 1951. 'Values and Value-Orientations in the Theory of Action: An Exploration in Definition and Classification'. In *Toward a General Theory of Action*, edited by T. Parsons and E. Shils, 388–433. Cambridge, MA: Harvard University Press.

Koepsell, David. 2009. *Who Owns You? The Corporate Gold Rush to Patent Your Genes*. Oxford: Wiley-Blackwell.

Krementsov, Nikolai. 2011. *A Martian Stranded on Earth: Alexander Bogdanov, Blood Transfusions, and Proletarian Science*. Chicago, IL: University of Chicago Press.
Kurzweil, Ray. 2005. *The Singularity Is Near: When Humans Transcend Biology*. New York: Viking.
———. 2003. 'Human Body Version 2.0'. *Lifeboat Foundation*. https://lifeboat.com/ex/human.body.version.2.0
———. 1999. *The Age of Spiritual Machines: When Computers Exceed Human Intelligence*. New York: Viking.
Laertius, Diogenes. 1961. 'Life of Zeno [of Critium]'. In *Essential Works of Stoicism*, edited by Moses Hadas, 1–47. New York: Bantam Books.
Lamb, David. 1988. *Down the Slippery Slope: Arguing in Applied Ethics*. New York: Croom Helm.
Lane, Robert E. 1966. 'The Decline of Politics and Ideology in a Knowledgeable Society'. *American Sociological Review* 31 (5): 649–62.
Langer, Susanne K. 1953. *Feeling and Form: A Theory of Art Developed from Philosophy in a New Key*. New York: Charles Scribner's Sons.
LaPiere, Richard. 1959. *The Freudian Ethic*. New York: Duell, Sloan, and Pearce.
Lassman, Peter, Irving Velody and Hermínio Martins, eds. 1989. *Max Weber's 'Science as a Vocation'*. London: Unwin Hyman.
Lebkowsky, Jon. 1997. 'The Cyborganic Path'. *CMC Magazine*, April. http://www.december.com/cmc/mag/1997/apr/lebkow.html.
Lecky, William Edward Hartpole. 1869. *History of European Morals from Augustus to Charlemagne*. London: Arno Press.
Le Figaro économie. 2004. July 29.
Lepage, Henri. 1978. *Demain le capitalisme*. Paris: le Livre de poche, Pluriel.
Lévy-Bruhl, Lucien. 1923. *Primitive Mentality*. London: George Allen & Unwin.
Lévy, Pierre. 1994. *L'intelligence collective – pour une anthropologie du cyberspace*. Paris: La Découverte.
Lewis, David K. 1986. *On the Plurality of Worlds*. Oxford: Blackwell.
Leys, Colin. 2001. *Market-Driven Politics: Neoliberal Democracy and the Public Interest*. London: Verso.
Lifton, Robert Jay. 1993. *The Protean Self: Human Resilience in an Age of Fragmentation*. Chicago, IL: University of Chicago Press.
Lowenthal, Leo. 1984. *Literature and Mass Culture*. Vol. 1. London: Transaction Books.
Luttwak, Edward. 1998. *Turbo-Capitalism: Winners and Losers in the Global Economy*. London: Weidenfeld & Nicholson.
Macfie, Alec Lawrence. 1967. *The Individual in Society: Papers on Adam Smith*. London: George Allen & Unwin.
Machlup, Fritz. 1962. *The Production and Distribution of Knowledge in the United States*. Princeton, NJ: Princeton University Press.
Mackenzie, Donald A. 1981. *Statistics in Britain, 1865–1930: The Social Construction of Scientific Knowledge*. Edinburgh: Edinburgh University Press.
Macmurray, John. 1957. *The Self as Agent*. London: Faber & Faber.
Magee, Bryan and Martin Milligan. 1995. *On Blindness: Letters between Bryan Magee and Martin Milligan*. Oxford: Oxford University Press.
Martins, Hermínio. 2014. 'On the Status and Role of Instrumental Images in Contemporary Science: Some Epistemological Issues'. *Scientiae Studia* 12: 11–36.
———. 2011. *Experimentum Humanum: Civilização Tecnológica e Condição Humana*. Lisbon: Relógio D'Água.
———. 2008. 'Biologia e politica: eugenismos de ontem e de hoje'. In *Ciencia e ciadadania: homenagem a bento de Jesus Caraça*, edited by Luísa Schmidt and João Pina-Cabral, 165–215. Lisbon: Imprensa das Ciências Sociais.
———. 2005. 'The Metaphysics of Information: The Power and the Glory of Machinehood'. *Res-Publica* 1: 65–192.
———. 2004. 'The Marketisation of Universities and Some Cultural Contradictions of Academic Capitalism'. *Metacrítica, Revista de filosofia* 4.

———. 2002. 'The Liminality of Human Reproductive Cloning, or In Praise of the Wisdom of Repugnance'. *Interact-Revista electrónica do Centro de Estudos de Comunicação e Linguagens*, Universidade Nova de Lisboa 3.

———. 2000. 'Ciência, técnica e utopia' ['Science, Technology and Utopia']. *LER* 48: 63–70.

———. 1998. 'Risk, Uncertainty and Eschatology: Reflections on the Ongoing Experimentum Mundi'. *Episteme-Revista da Universidade Técnica de Lisboa* 1 (1): 99–121.

———. 1996. *Hegel, Texas e outros ensaios de teoria social [Hegel, Texas and Other Essays in Social Theory]*. Lisbon: Século XXI.

———. 1993. 'Hegel, Texas – Issues in the Sociology and Philosophy of Technology'. In *Knowledge and Passion: Essays in Honour of John Rex*, edited by Hermínio Martins, 226–49. London: I.B. Tauris.

———. 1983. 'Tristes durées'. In *Death in Portugal*, edited by R. Feijo, H. Martins and J. Pina-Cabral. Oxford: JASO.

———. 1972. 'The Kuhnian 'Revolution' and Its Implications for Sociology'. In *Imagination and Precision in the Social Sciences – Essays in Memory of Peter Nettl*, edited by Tom Nossiter, A. H. Hanson and Stein Rokkan, 13–58. London: Faber and Faber.

Marx, Leo. 1964. *The Machine in the Garden: Technology and the Pastoral Ideal in America*. Oxford: Oxford University Press.

Masing-Delic, Irene. 1992. *Abolishing Death: A Salvation Myth of Russian Twentieth-Century Literature*. Stanford: Stanford University Press.

Maslow, Abraham. 1954. *Motivation and Personality*. New York: Harper & Brothers.

Mauss, Marcel. 1985. 'A Category of the Human Mind'. In *The Category of the Person: Anthropolo gy: Philosophy, History*, edited by Michael Carrithers, Steven Collins and Steven Lukes, 1–25. Cambridge: Cambridge University Press.

———. 1934. 'Les techniques du corps' ['The Techniques of the Body']. *Journal de Psychologie* 32 (3–4).

McGee, Glenn. 1996. *The Perfect Baby: A Pragmatic Approach to Genetics*. Lanham, MD: Rowman & Littlefield.

McGuire, James E. and Piyo M. Rattansi. 2007. 'Newton Y Las 'Flautas De Pan'. *Estud.filos* 35: 149–87. Medellín: Universidad de Antioquia.

McHenry, Leemon B. 1992. *Whitehead and Bradley: A Comparative Analysis*. Albany: State University of New York.

McNeill, William H. 1986. 'Mythistory, or Truth, Myth, History, and Historians'. *The American Historical Review* 91 (1): 1–10.

Mécary, Caroline. 2013. 'GPA: la plus grande subversion féministe'. *Le Monde*, 2 July. http://www.lemonde.fr/idees/article/2013/02/07/gpa-la-plus-grande-subversion-feministe_1828552_3232.html. Accessed 19 July 2018.

———. 2012. *L'Amour et la Loi*. Paris: Alma Editeur.

Merton, Robert K. 1968. 'The Matthew Effect in Science: The Reward and Communication Systems of Science Are Considered'. *Science* 159 (3810): 56–63.

Meyer, John W. et al. 1977. 'The World Educational Revolution, 1950–1970'. *Sociology of Education* 50 (4): 242–58.

Michalson, Gordon E. Jr. 1990. *Fallen Freedom: Kant on Radical Evil and Moral Regeneration*. Cambridge: Cambridge University Press.

Miller, Paul and James Wilsdon, eds. 2006. *Better Humans? The Politics of Human Enhancement and Life Extension*. London: Demos.

Mondolfo, Rodolfo. 1952. *El infinito en el pensamiento de la antigüedad clásica*. Buenos Aires: Imán.

Montague, William Pepperell. 1925. *The Ways of Knowing or the Methods of Philosophy*. New York: Macmillan.

Morris, Charles W. 1964. *Signification and Significance: A Study of the Relations of Signs and Values*. Cambridge, MA: MIT Press.

Moscovici, Serge. 1981. *L' Âge des Foules: Un traité historique de psychologie des masses*. Paris: Fayard.

Muller, H. J. 1935. *Out of the Night: A Biologist's View of the Future.* New York: Vanguard Press.
Mundy, Liza. 2007. *Everything Conceivable: How Assisted Reproduction Is Changing Men, Women and the World.* London: Knopf.
Münster, Arno. 2002. *Heidegger, la 'science allemande' et le national socialisme.* Paris: Kimé.
Musso, Pierre. 1987. 'Aux origines du concept moderne: corps et réseau dans la philosophie de Saint Simon'. *Année* 3 (1): 11–29.
Naam, Ramez. 2005. *More than Human: Embracing the Promise of Biological Enhancement.* New York: Broadway Books.
New Scientist. 2007. 13 October.
Nozick, Robert. 1974. *Anarchy, State, and Utopia.* New York: Basic Books.
Nuland, Sherwin. 2005. 'Do You Want to Live Forever?' *MIT Technology Review*, 1 February. https://www.technologyreview.com/s/403654/do-you-want-to-live-forever/. Accessed 19 July 2018.
Oakeshott, Michael. 1959. *The Voice of Poetry in the Conversation of Mankind: An Essay.* London: Bowes & Bowes.
Osborne, Peter. 2000. *Philosophy in Cultural Theory.* London: Routledge.
Ossowski, Stanislaw. 1963. *Class Structure in the Social Consciousness.* New York: Free Press of Glencoe.
Palumbi, Stephen R. 2001. *The Evolution Explosion: How Humans Cause Rapid Evolutionary Change.* New York: Norton.
Panofsky, Wolfgang K. H. 2007. *Panofsky on Physics, Politics, and Peace: Pief Remembers*, edited by Jean Marie Deken. New York: Springer Science.
Parsons, Talcott. 1950. 'The Prospects of Sociological Theory'. *American Sociological Review* 15 (1): 3–16.
Parsons, Talcott, George M. Platt and Neil J. Smelser (contributor). 1973. *The American University.* Cambridge, MA: Harvard University Press.
Passmore, John. 1970. *The Perfectibility of Man.* London: Duckworth.
Perry, Ralph Barton. 1910. 'The Egocentric Predicament'. *Journal of Philosophy, Psychology and Scientific Methods* 7: 5–14.
Petryna, Adriana. 2009. *When Experiments Travel: Clinical Trials and the Global Search for Human Subjects.* Princeton, NJ: Princeton University Press.
Pichot, André. 2000. *La société pure, de Darwin à Hitler.* Paris: Flammarion.
Pickering, W. S. F. and Hermínio Martins, eds. 1994. *Debating Durkheim.* London: Routledge.
Pigou, Arthur. 1920. *The Economics of Welfare.* London: Macmillan.
Pind, Jörgen L. 2014. *Edgar Rubin and Psychology in Denmark: Figure and Ground.* New York: Springer International.
Popper, Karl. 1999. [1994]. *All Life Is Problem Solving.* New York: Routledge.
———. 1957. *The Poverty of Historicism.* New York: Routledge.
———. 1945. *The Open Society and Its Enemies.* London: Routledge.
Porat, Marc U. 1977. *The Information Economy: Definition and Measurement.* Washington, DC: Office of Telecommunications.
Posner, Richard A. 1992. *Sex and Reason.* Cambridge, MA: Harvard University Press.
Ramsey, Paul. 1970. *Fabricated Man: The Ethics of Genetic Control.* New Haven, CT: Yale University Press.
Rawls, John. 1971. *A Theory of Justice.* Cambridge: Belknap Press.
Readings, Bill. 1996. *The University in Ruins.* Cambridge, MA: Harvard University Press.
Rees, Martin. 2003. *Our Final Hour*, a *Scientist's Warning: How Terror, Error, and Environmental Disaster Threaten Humankind's Future in This Century – On Earth and Beyond.* New York: Basic Books.
Reichenbach, Hans. 1938. *Experience and Prediction: An Analysis of the Foundations and the Structure of Knowledge.* Chicago, IL: University of Chicago Press.
Renouvier, Charles Bernard. 1854. *Essais de critique générale.* Paris: Librairie Philosophique de Ladrange.
Rescher, Nicholas. 1969. *Essays in Philosophical Analysis.* Pittsburgh, PA: University of Pittsburgh Press.
Rieff, Philip. 1959. *Freud: The Mind of the Moralist.* New York: Anchor Books.

Riesman, David, Nathan Glazer and Reuel Denney. 1950. *The Lonely Crowd: A Study of the Changing American Character*. New Haven, CT: Yale University Press.

Rifkin, Jeremy. 1987. *Time Wars: The Primary Conflict in Human History*. New York: Henry Holt.

———. 1983. *Algeny: A New Word – A New World*. New York: Viking.

Roff, Sue Rabbitt. 2011. 'We Should Consider Paying Kidney Donors'. *The British Medical Journal* 2 August. http://www.bmj.com/content/343/bmj.d4867. Accessed 19 July 2018.

Rolo, José Manuel. 2006. *O regresso às armas – tendências das indústrias da defesa*, Lisbon: Edições Cosmos.

Rothblatt, Martine. 1997. *Unzipped Genes: Taking Charge of Baby-Making in the New Millennium*. Philadelphia, PA: Temple University Press.

Russell, Bertrand. 2009 [1931]. *The Scientific Outlook*. Routledge: London.

———. 1940. *An Inquiry into Meaning and Truth*. New York: W. W. Norton.

Ryle, Gilbert. 1949. *The Concept of Mind*. Chicago, IL: University of Chicago Press.

Satel, Sally. 2011. 'The Market for Kidneys, Livers and Lungs'. *Wall Street Journal*, November 8. https://www.wsj.com/articles/SB10001424052970204554204577023623689583052.

Sartre, Jean-Paul. 1946. *L'existentialisme est un humanisme*. Paris: Nagel.

Schnaiberg, Allan. 1980. *The Environment: From Surplus to Scarcity*. New York: Oxford University Press.

Schoeck, Helmut. 1969 [1966]. *Envy: A Theory of Social Behaviour*. New York: Harcourt, Brace & World.

Schumpeter, Joseph A. 1942. *Capitalism, Socialism, and Democracy*. New York: Harper.

Sfez, Lucien. 1995. *La santé parfait: Critique d'une nouvelle utopie*. Paris: Seuil.

Shaw, George Bernard. 1934 [1933]. 'On the Rocks'. In *Too True to Be Good, Village Wooing & On the Rocks*. London: Constable.

Shelley, Mary. 1818. *Frankenstein; or the Modern Prometheus*. London: Lackington, Hughes, Harding, Mavor & Jones.

Shostak, Stanley. 2002. *Becoming Immortal: Combining Cloning and Stem-Cell Therapy*. Albany: State University of New York Press.

Sills, David L. 1957. *The Volunteers: Means and Ends in a National Organization*. Glencoe: The Free Press.

Silver, Lee M. 1997. *Remaking Eden: How Genetic Engineering and Cloning Will Transform the American Family*. New York: Avon Books.

Simmel, Georg. 1900. *The Philosophy of Money* [*Philosophie des Geldes*]. Leipzig: Duncker & Humblot.

Sinclair, Upton. 1923. *The Goose-Step: A Study of American Education*. Pasadena: Published by author.

Skolimowski, Henryk. 1967. *Polish Analytical Philosophy: A Survey and a Comparison with British Analytical Philosophy*. London: Routledge & Kegan Paul.

Smil, Vaclav. 2006. *Transforming the Twentieth Century: Technical Innovations and Their Consequences*. Oxford: Oxford University Press.

Smith, Wesley J. 2004. *Consumer's Guide to a Brave New World*. New York: Encounter Books.

Sober, Elliott and David Sloan Wilson. 1998. *Unto Others: The Evolution and Psychology of Unselfish Behavior*. Cambridge, MA: Harvard University Press.

Sorokin, Pitirim A. 2010. *Man and Society in Calamity*. Reprint edition. New Brunswick: Transaction.

Spar, Debora L. 2006. *The Baby Business: How Money, Science, and Politics Drive the Commerce of Conception*. Boston, MA: Harvard Business School Press.

Spencer, Herbert. 1950 [1857]. *Literary Style and Music*. London: Watts.

———. 1857. 'Progress: Its Law and Cause'. *Westminster Review* 67: 445–85.

Spolsky, Ellen. 1993. *Gaps in Nature: Literary Interpretation and the Modular Mind*. Albany: State University of New York Press.

Sprigge, Timothy. 1993. *James and Bradley: American Truth and British Reality*. Chicago, IL: Open Court.

Stivers, Richard. 1999. *Technology as Magic: The Triumph of the Irrational*. New York: Continuum.

Stock, Gregory. 2002. *Redesigning Humans: Our Inevitable Genetic Future*. New York: Houghton Mifflin.

———. 1993. *Metaman: The Merging of Humans and Machines into a Global Superorganism.* New York: Simon & Schuster.

Stonier, Tom. 1997. *Information and Meaning: An Evolutionary Perspective.* London: Springer-Verlag.

———. 1990. *Information and the Internal Structure of the Universe: An Exploration into Information Physics.* London: Springer-Verlag.

Strawson, P. F. 1959. *Individuals: An Essay in Descriptive Metaphysics.* London: Methuen.

Sunstein, Cass R. and Richard H. Thaler. 2003. 'Libertarian Paternalism Is Not an Oxymoron'. *The University of Chicago Law Review* 70 (4): 1159–202.

Sylvan, Richard and David Bennett. 1994. *The Greening of Ethics: From Anthropoceneto Deep-Green Theory.* Tucson: University of Arizona Press.

Tabarrok, Alexander, ed. 2002. *Entrepreneurial Economics: Bright Ideas from the Dismal Science.* Oxford: Oxford University Press.

Taylor, Gordon Rattray. 1968. *The Biological Time Bomb.* London: Thames & Hudson.

———. 1967. *The Science of Life: A Pictorial History of Biology.* London: Panther.

Taylor, James Stacey. 2005. *Stakes and Kidneys: Why Markets in Human Body Parts Are Morally Imperative.* New York: Routledge.

Teilhard de Chardin, Pierre. 1955. *Le phénomène humain.* Paris: Éditions du Seuil.

Thaler, Richard H. and Cass R. Sunstein. 2008. *Nudge: Improving Decisions about Health, Wealth, and Happiness.* New Haven, CT: Yale University Press.

Theweleit, Klaus. 1987. *Male Fantasies.* Vol. 1. Minneapolis: University of Minnesota Press.

Thomas, Keith. 1971. *Religion and the Decline of Magic.* New York: Scribner.

Thomas, Lewis. 1992. *The Fragile Species.* New York: Charles Scribner's Sons.

Thompson, Charis. 2005. *Making Parents: The Ontological Choreography of Reproductive Technologies.* Cambridge, MA: MIT Press.

Times. 2002. January 11.

Tipler, Frank J. 1994. *The Physics of Immortality: Modern Cosmology, God and the Resurrection of the Dead.* New York: Doubleday.

Titmuss, Richard. 1970. *The Gift Relationship: From Human Blood to Social Policy.* London: George Allen & Unwin.

Toulmin, Stephen. 1990. *Cosmopolis: The Hidden Agenda of Modernity.* Chicago, IL: University of Chicago Press.

Toye, John. 2000. *Keynes on Population.* New York: Oxford University Press.

Tucker, Robert C. 1961. *Philosophy and Myth in Karl Marx.* Cambridge: Cambridge University Press.

Veblen, Thorstein. 1918. *The Higher Learning in America: A Memorandum on The Conduct of Universities By Business Men.* New York: B. W. Huebsch.

Wade, Nicholas. 2009. 'Quest for a Long Life Gains Scientific Respect'. *New York Times*, 28 September. http://www.nytimes.com/2009/09/29/science/29aging.html?mcubz=1.

———. 2001. *Life Script: How the Human Genome Discoveries Will Transform Medicine and Enhance Your Health.* New York: Simon & Schuster.

Wakefield, Andre. 2010. 'Leibniz and the Wind Machines'. *Osiris* 25 (1).

Ward, Peter D. and Donald Brownlee. 2000. *Rare Earth: Why Complex Life Is Uncommon in the Universe.* New York: Copernicus Books.

Warnock, Mary. 2002. 'Genetic Engineering and What Is Natural'. *Think* 1 (1): 21–27.

Warwick, Kevin. 2002. *I, Cyborg.* London: Century.

Wasserstrom, Jeffrey. 2012. 'Introduction: Who Are You This Time?'. In *Chinese Characters: Profiles of Fast-*Changing *Lives in a Fast-Changing Land*, edited by Angilee Shah and Jeffrey Wasserstrom, 1–12. Berkeley: University of California Press.

Watkins, John W. N. 1984. *Science and Scepticism.* Princeton, NJ: Princeton University Press.

Weber, Max. 1930. *The Protestant Ethic and the Spirit of Capitalism*, translated by Talcott Parsons. London: George Allen & Unwin.

Weismann, August. 1893. *The Germ-Plasm: A Theory of Heredity.* New York: Charles Scribner's Sons.

Wertheim, Margaret. 1999. *The Pearly Gates of Cyberspace: A History of Space from Dante to the Internet.* New York: W. W. Norton.

Whitehead, Alfred North. 1928. *Symbolism: Its Meaning and Effect.* Cambridge: Cambridge University Press.

Whyte, William H. 1956. *The Organization Man.* New York: Simon & Schuster.

Wiles, Peter. 1962. *The Political Economy of Communism.* Oxford: B. Blackwell.

Wilson, Logan. 1942. *The Academic Man: A Study in the Sociology of a Profession.* London: Oxford University Press.

Wittgenstein, Ludwig. 1953. *Philosophical Investigations*, translated by G. E. M. Anscombe. Oxford: B. Blackwell.

Wolin, Richard. 2001. *Heidegger's Children: Hannah Arendt, Karl Löwith, Hans Jonas, and Herbert Marcuse.* Princeton, NJ: Princeton University Press.

INDEX

Lightning Source UK Ltd.
Milton Keynes UK
UKHW040754201218
334276UK00001B/64/P

9 781783 088324